Automation in Virtual Testing of Mechanical Systems

Automation in Virtual Testing of Mechanical Systems

Theories and Implementation Techniques

Ole Ivar Sivertsen

Bjørn Haugen

CRC Press
Taylor & Francis Group
Boca Raton London New York

CRC Press is an imprint of the
Taylor & Francis Group, an **informa** business

CRC Press
Taylor & Francis Group
6000 Broken Sound Parkway NW, Suite 300
Boca Raton, FL 33487-2742

© 2019 by Taylor & Francis Group, LLC
CRC Press is an imprint of Taylor & Francis Group, an Informa business

No claim to original U.S. Government works

Printed on acid-free paper
Version Date: 20180606

International Standard Book Number-13: 978-1-1386-1076-7 (Hardback)

Library of Congress Cataloging-in-Publication Data

Names: Sivertsen, Ole Ivar, 1945- author. | Haugen, Bjørn, 1963- author.
Title: Automation in virtual testing of mechanical systems / Ole Ivar
  Sivertsen and Bjørn Haugen.
Description: First edition. | Boca Raton, FL : CRC Press/Taylor & Francis
  Group, LLC, 2018. | "A CRC title, part of the Taylor & Francis imprint, a
  member of the Taylor & Francis Group, the academic division of T&F Informa
  plc." | Includes bibliographical references.
Identifiers: LCCN 2018009215| ISBN 9781138610767 (hardback : acid-free paper)
  | ISBN 9781138032590 (e-book)
Subjects: LCSH: Machinery--Testing--Data processing. | Machinery--Computer
  simulation.
Classification: LCC TJ148 .S575 2018 | DDC 621.8028/7--dc23
LC record available at https://lccn.loc.gov/2018009215

Visit the Taylor & Francis Web site at
http://www.taylorandfrancis.com

and the CRC Press Web site at
http://www.crcpress.com

Our different farming experiences at Skjerstad and Lånke

Contents

Preface   xi

Acknowledgment   xiii

Authors   xv

List of Abbreviations   xvii

1 Introduction   **1**
  1.1 About Knowledge-Based Engineering (KBE)   2
  1.2 About Dynamic Simulation   3
  1.3 About Optimization Techniques   4
  1.4 The Structure of the Book   5

2 State-of-the-Art Knowledge-Based Engineering   **7**
  2.1 Definitions and Classifications   8
  2.2 Features of KBE Languages   10
  2.3 Knowledge Acquisition for KBE   12
    2.3.1 Tools for knowledge acquisition and modeling   16
  2.4 Developing KBE Applications   23
  2.5 Knowledge Acquisition for Introducing KBE in Mechanism De-
     sign   24

3 State-of-the-Art FE-Based Dynamic Simulation   **29**
  3.1 Description of Motion   29
    3.1.1 Notation   29
    3.1.2 Rigid body motion   29
      3.1.2.1 Rotation   31
      3.1.2.2 Rodrigues parameterization of rotations . . .   31
    3.1.3 Variation of Rodrigues parameterization   32
  3.2 Finite Element Theory   32
  3.3 Corotational Formulation   33
  3.4 Mechanism Links as Substructures and Super Elements . . .   34
    3.4.1 Static modes   34
    3.4.2 Fixed interface dynamic modes   35
    3.4.3 Reduced system   35
    3.4.4 Structural damping   37

3.5   FE Modeling of Joints   39
    3.5.1   Description of joint motion   39
        3.5.1.1   Variation of joint motions   39
3.6   Joint Specializations   40
    3.6.1   Rigid joint   40
    3.6.2   Revolute, cylindric and prismatic joint with a single
        master   40
    3.6.3   Universal joint   41
    3.6.4   Constant velocity joint   41
    3.6.5   Multimaster joints   42
    3.6.6   Transmissions between joint variables   43
3.7   Multipoint Constraints   44
    3.7.1   Rigid multipoint constraint   44
    3.7.2   Weighted average multipoint constraint   45
3.8   Time Integration Methods for Nonlinear Dynamics   47
    3.8.1   Newmark integration   48
        3.8.1.1   Newmark with respect to displacement incre-
            ment   48
    3.8.2   Generalized α method   49
    3.8.3   Generalized α method with Newmark integration . . .   50
        3.8.3.1   Equilibrium iterations; Newton iterations . .   51
        3.8.3.2   Stability and accuracy   52

4   Framework for Generic Mechanism Modeling   55
4.1   The Sheth–Uicker (SU) Formulation   55
4.2   Generic Library Format for Mechanisms   56
    4.2.1   Node positions   56
    4.2.2   Constraints   56
    4.2.3   Link shapes   58
    4.2.4   Springs and dampers   59
    4.2.5   Loads   60
4.3   Default Link and Joint Shapes   60
    4.3.1   Connection shape   61
    4.3.2   Sweep   61
    4.3.3   Surfaces   62
    4.3.4   Joint geometry   63
    4.3.5   Assembly   63
4.4   Extended Modeling of Link Shapes   64
4.5   Automated Generation of Simulation Input   64
    4.5.1   Finite element data for mechanism links   64
    4.5.2   Mechanism system data   65
4.6   Example Mechanism with Different Modeling Elements   . . .   66
    4.6.1   Example joint definition input   66
    4.6.2   Example link shape definition input   67
    4.6.3   Details about FE meshing of links and joints   71

	4.6.4	Importing demonstration example into FEDEM	74

5 Design Optimization — **77**

5.1	Formulation of Optimization Problems	77
	5.1.1	Design variables	77
	5.1.2	Objective function	78
	5.1.3	Constraints	78
	5.1.4	A standard problem	79
	5.1.5	Fundamental concepts	79
5.2	Optimization Methods	81
	5.2.1	Methods for unconstrained optimization	81
	5.2.2	Methods for constrained optimization	82
	5.2.3	Global optimization methods	83
5.3	Multidisciplinary Design Optimization	83
5.4	Multiobjective Optimization	84
5.5	Optimization of Dynamic Performance	86
5.6	Optimizing Flexible Multibody Systems	87
	5.6.1	Evaluating sensitivities, objectives and constraints . .	87
	5.6.2	Surrogate models and response surfaces	89
5.7	Tool for Optimizing Flexible Multibody Systems	90

6 Environment for Design Automation — **93**

6.1	KBE Development Framework	93
6.2	Boolean Geometry	95
6.3	Automatic Meshing	96

7 Interaction with the User — **101**

7.1	Entering and Editing the Mechanism Library Format	101
	7.1.1	Menus for entering/editing the KBE library format . .	103
	7.1.2	Link editing module	106
7.2	Other Simulation Input	108
7.3	Design Parameters and Optimization	110
7.4	Extended KBE Programming	112

8 Automated Design Cases — **113**

8.1	Linkages	113
	8.1.1	Four-bar mechanisms	114
	8.1.2	Six-link mechanisms	116
8.2	Suspension Systems	120
	8.2.1	Double wishbone suspension	125
8.3	Cranes and Robots	132
	8.3.1	Offshore crane	133
8.4	Wind Turbines	138
	8.4.1	An offshore wind turbine model	138
8.5	Optimization for Automation in Mechanism Design	146

8.5.1   Optimization approaches demonstrated on a four-bar
        mechanism   146
        8.5.1.1   Optimization without constraints   148
        8.5.1.2   Optimization with constraint   148
        8.5.1.3   Optimization on surrogate model   151
        8.5.1.4   Discussion of case results   153
8.5.2   Optimization approaches demonstrated on a Stephen-
        son 3 mechanism   153
        8.5.2.1   Parameterization   154
        8.5.2.2   Discussion of case results   158

9 Discussions and Conclusions   161
9.1   Modeling of Planar and Spatial Mechanisms   161
9.2   Effective User Interaction   163
9.3   The Optimization Loop   164
9.4   Conclusions   165

References   167

Index   173

Preface

Mechanical engineering is probably the most basic engineering discipline and started several thousand years back, but is still the backbone of any engineering activity today. Mechanisms and mechanism design are almost synonymous with mechanical engineering and are always key when designing devices that should control motion in any way be it a crane, a vehicle, an airplane and many other both simple and complex systems. Designing a mechanism, for instance a four-bar mechanism could in principle be very simple and intuitive because it is very visual and that is maybe the reason that these kinds of devices were made thousands of years ago. The design process would be to move hinges around in a physical model and by inspection see the behavior of the system and repeatedly make changes and inspect the behavior until satisfactory operation was confirmed. When more complex and precise systems were needed, mathematical tools like synthesis and analysis were required. Today a large number of such software tools are available and one such tool was presented in the precursor for this book [51], a comprehensive simulation tool for flexible multibody systems (FMBS) [3].

Acknowledgment

The support from the TechnoSoft Company has been a prerequisite for developing the KBE pilot for mechanism design and we are very grateful to the head of the company, Adel Chemaly, and to a number of his employees for their support and help during the stay in Ohio with the first KBE pilot implementation and later supporting master theses at Norwegian University of Science and Technology extending this implementation.

Fedem [3] is a commercial software package to simulate mechanisms and flexible multibody systems. From 2014 to 2016, the KBE application for mechanism simulation in Fedem has been developed as master thesis projects by Rasmus Korvald Skare [52], Anders Kristiansen and Eivind Kristoffersen [24], Thor Christian Coward [10] and Arnt Underhaug Lima [25].

In the early 1990s, the cooperation with Dr. Bernhard Specht [57, 56] laid a foundation for optimization in mechanism simulation during a project financed by the European Union. This paved the way for the doctoral thesis by Sigurd Due Trier [61] in 2001, and he has written large portions of Sections 5.1, 5.2, 5.5 and 5.6.

Professor Terje Rølvåg gave some very valuable help setting up some realistic simulations in Chapter 8, drawing on his extensive experience with dynamic simulation.

Without the combined effort from the people referred to above this book would not have been possible and we are very grateful to these people for their contributions.

Ole Ivar Sivertsen
Bjørn Haugen

Authors

Ole Ivar Sivertsen is Professor of Mechanical Engineering at Norwegian University of Science and Technology in Trondheim, Norway. He was department head 1996–98 and 2007–12 and study program head 2004–13. He was originator of the simulation software FEDEM developed in national and international research projects and commercialized in the 1990s through the Fedem Technology Company. It has applications in automotive, aerospace, robotics, tractors, construction
machines, offshore structures, machine tools, etc. The simulations are based on structural dynamics using FE and control engineering for simulation of strength and product behavior. He is author of the book *Virtual Testing of Mechanical Systems, Theories and Techniques*, 2001. In recent years his research focus has been knowledge-based engineering (KBE).

Bjørn Haugen is an associate professor of mechanical engineering at Norwegian University of Science and Technology in Trondheim, Norway. His educational background is civil and aerospace engineering sciences. His research topics are finite element methods with applications to structural mechanics, elastic mechanisms and tribology. His industry experience is in making engineering analysis software at Fedem Technolgy Inc. and biomedical software at HYTEC Inc.

List of Abbreviations

AI	Artificial Intelligence
AML	Adaptive Modeling Language
ASCII	American Standard Code for Information Interchange
BDF	NASTRAN Bulk Data Format
CAD	Computer Aided Design
CDM	Critical Decision Method
CBR	Case-Based Reasoning
CLOS	Common Lisp Objective System
CMS	Component Mode Synthesis
DOF	Degrees of Freedom
FE	Finite Elements
FMBS	Flexible MultiBody Systems
GA	Genetic Algorithms
GDL	General-purpose Declarative Language of Genworks
HHT	Hilber–Hughes–Taylor
ICARE	Illustrations, Constraints, Activities, Rules and Entities (Refer MOKA)
IDL	ICAD Design Language
IGES	Initial Graphics Exchange Specification
KA	Knowledge Acquisition
KBE	Knowledge-Based Engineering
K-briefs	A3 format Knowledge Briefs
MDO	Multidisciplinary Design Optimization
MML	MOKA Modeling Language
MOKA	Methodology and Tools Oriented to Knowledge-Based Engineering Applications
MPC	Multipoint Constraints
NURBS	Non-Uniform Rational B-Spline
OPM	Object Process Methodology
OO	Object Oriented
PTO	Power Take Off
RaMMS	Rapid Mechanism Modelling System
RBE2	Rigid Multipoint Constraints
RBR	Rule-Based Reasoning
RIF	Rule Interchange Format
RS	Response Surfaces

S-LDO	Single-Level Design Optimizations
SM	Surrogate Models
STEP	Standard for the Exchange of Product Model Data
SQP	Sequential Quadratic Programming
STL	Stereolithography
SU	Sheth–Uicker Formulation
UML	Unified Modeling Language
XML	Extensible Markup Language

1

Introduction

Design of machine systems could be a very demanding task both with respect to complexity and resources and because many engineering and non-engineering disciplines are often involved. Traditionally machine systems may be assembled from subassemblies in several levels all the way to bolts and nuts. A machine will almost always have constrained motion between moving parts, be it between crankshaft and pistons in an engine, between parts in a suspension system for a car or an aircraft, or between the parts of handling equipment like a linkage or a robot. These systems with constrained motion are named mechanisms where the moving parts are called links and the motion of the links is constrained by joints that couple the links together. A large set of joint types is available for constraining link relative motion (for more details refer to comprehensive literature on mechanisms and mechanism joints). Other issues that need to be considered when designing machines are cost analysis, environmental issues, electronics and instrumentation and, last but not least, aesthetics, especially if the machine is a consumer product and to a lesser extent if it is a production tool.

Mechanisms vary in a close to unlimited number of topologies and dimensions, and designing a mechanism manually could be very time consuming, especially if you need to do structural analysis and simulations as part of the design loop. This book will try to outline ways to automate significant parts of this design loop for mechanisms by utilizing knowledge-based engineering (KBE) to generate parameterized geometries for links and joints and from this generate input to dynamic or static simulations like positions of joints, topologies, link geometries, forces, motion input and FE meshes for links and joints. The tedious job of making simulation input can then be automated, drastically reducing the risk of making errors from manually entering the input data to the simulation.

Dynamic or static simulation is a way to evaluate a proposed design both with respect to structural integrity, and also with respect to the functional requirements of the system. Dynamic simulation could generate large amounts of output data that could be visualized as curve plots and motion visualization including stress contour plots.

Following the simulations in the automated design loop could be a spot where the designer likes to interrupt the design loop to inspect the progression of the results visually, for instance as animations with stress contour plots. The designer should be able to specify where to interrupt the automated design loop. To utilize automated optimization in the design loop, a set of optimization algorithms suited for this type of optimization is needed, and in addition the designer needs to define an objective function based on the requirements to be able to generate updated design input based on the simulation results as an alternative to manually inspecting the simulation results and from that specify updated simulation input.

Today's trend for products is towards custom-made products and smaller production series, and this requires increased design work including repeated simulations on large variations within the same product family. Automating design work in this setting will give large cost and time savings that could be critical in a very competitive market. Quality gains from standardization and structuring the design information will give additional benefits.

1.1   About Knowledge-Based Engineering (KBE)

The term knowledge-based engineering (KBE) is not very useful for communicating what the technology is about. If someone working in KBE explains to another engineer or designer what she is doing using the term "knowledge-based engineering", she has a good chance of getting the response one of the authors received: "What do you think I am doing - Engineering NOT based on knowledge?" For easier communication some in the KBE community have started to use the term "automation in design", which is not a very precise term, but explains much more plainly what KBE is about.

Most engineers and designers know very well what the term "computer aided design" (CAD) is about. Briefly one could explain CAD as an interactive graphic tool used to enter the engineer's idea of a geometry into the computer. For the more advanced CAD systems you are also able to parameterize the geometry; so by changing values for parameters you can change dimensions and form of the geometric model. What is typical for CAD tools is that the process of building the geometry is manual based on some ideas in the designer's head and possibly supported by some handmade sketches.

Using KBE, you also often will produce what could be called CAD geometry. However, the geometry is not built interactively by the designer, but generated by a computer program, thereby giving much more flexibility for variation based on some rules implemented in the program. In a CAD system you could make a part and reuse it in several positions by just translating and rotating from the design position (refer to, for instance, the legs of a table), but this positioning is also done manually by the designer, and possibly parameterized. Doing this in a KBE application, the whole process is automated in a program, but in addition the different parts that are positioned could be generated with different shapes and details.

Using our case of mechanism design as an example, the aim is to generate a mechanism with a number of links connected with joints of possibly different types, for instance selected from a joint library. Each link could be very different from other links both regarding the number of joints connected to the link and joint types. The KBE application will based on some simple generic input specifications for a mechanism generate all the mechanisms links including detailed joint geometries based on an algorithm for link generation in the KBE application. The aim of our KBE application is to generate input data for simulation of the dynamics of flexible mechanisms. Input data such as loading, springs and dampers, sensors and actuators, control elements, FE meshes for the links, etc., must be generated by the KBE application. More details about knowledge based engineering appear in Chapter 2.

1.2 About Dynamic Simulation

The term dynamic simulation is a very broad term, but in this context we will limit ourselves to dynamic simulation of mechanisms based on a non linear finite element formulation. A mechanism is defined as a mechanical system consisting of links that may have spatial motion constrained by a variety of joint types like revolute joints, ball joints, sliders and many other joints. To build a simulation model you need to define the geometry for every link and joint in the system to be simulated and put it together in a design position for the system. To model the elasticity of the system finite element (FE) meshes have to be generated for every link including connection points for every joint referred to. In addition to the mechanical structures as mentioned above, you need to specify motion inputs like loading and prescribed motion, for instance as loads and torques, but also elements like springs and dampers are very often needed to define the system. Springs and dampers may be between points on links, but are even more likely to be defined in joints. Control systems are very often needed as part of the simulations to produce input to actuators in the system in the form of hydraulic cylinders and electric motors and with input to the control system from sensors and measurements in the system. A comprehensive simulation model is used to evaluate the mechanical structure of the system with respect to mechanical integrity, but also with respect to functionality requirements for the system.

Results from dynamic simulations are first of all motions for the different parts of the system like positions, velocities and accelerations. This may be presented as curves for different positions in the system like coupler curves that shows how a certain point moves during system operation, but also curves for velocities and accelerations. This type of curve may also be produced for joint variables and for forces and torques produced in the system. The simulation results are produced by dynamic integration of the equations generated from the simulation model. If detailed deformations and stresses are needed, for instance for evaluating stress levels with respect to material strength limits, a post processing is needed to retrack these deformations and stresses where required based on the simulation results from the time integration.

Building a simulation model as referred to above could be quite tedious and the possibilities for introducing errors are large. If it was just one single part to be modeled it could also be parameterized in a CAD system and re-meshed automatically. However, for a system of bodies connected by joints and where a design option could be to move the joints around in the design position, using traditional CAD systems will require a lot of manual work to update the geometry and the simulation model. This will mean extra cost and time and also likely introduce errors to be corrected by test simulations. This is where KBE comes in: If you could develop a KBE system that, based on engineering rules, could change the geometry from global changes and generate new link and joint geometries automatically, new FE meshes for the links and an updated simulation model could also be generated automatically and a new simulation could be started with no or minimum manual work. This could speed up the design loop significantly.

In the KBE application described above, the global design modifications may be done manually by the designer by changing a few design parameters, for instance a global position of a joint. In many cases this may be the best solution, at least for the first design iterations, for design modification because it could be difficult to formulate the design intent into a robust objective function. This will be discussed in more in detail the chapter about optimization, Chapter 5. However, there are situations where the designer may use trial and error for design modifications more or less by chance and this may lead to a lot of manual iterations. If an objective function could be formulated, these iterations could be automated using optimization techniques combined with automatically repeated simulations. More details about dynamic simulation appear in Chapter 3.

1.3   About Optimization Techniques

Optimization techniques could be defined as a technology used to search for an optimal solution for a problem based on mathematically formulated requirements: The objective function. Running optimization means searching for a minimum point for the objective function where each evaluation represents one point on for instance a surface. In addition to searching such a surface for a minimum there may be constraints limiting the search area. A picture that could visualize the optimization process could be searching for the lowest point on a bumpy farm surrounded by fences limiting the search area, the constraints. However, finding a minimum point does not guarantee that this is the optimum solution. The solution found could be a local optimum and further search will be needed to find a potential global optimum. As we know, the first derivatives for a surface in a minimum point will be zero, however, a constraint may require that the optimum solution is not in a point with the derivative equal to zero, but limited by a constraint on the slope towards the minimum point on the surface.

Calculations in each optimization step may be as easy as calculating the volume of a box, but could also be a very demanding computational task such as running a dynamic simulation of a complex system. The optimization approach may be very different depending on how consuming these computations are. If the evaluation of the response is very fast you can afford to evaluate many points on the surface to localize a point that has the minimum value. However, if time consuming simulations are needed in every optimization step, an approach is needed that requires few repeated simulations. One approach could be to calculate sensitivities for the design variables to be able to select the best combination of changes in the design variables to move in the direction on the surface towards the minimum point. This will speed up the process of finding the minimum point. Calculating these sensitivities is not usually done by simulation codes and sensitivities may therefore not be available for the optimization process.

An alternative approach to limit the number of simulations for an optimization could be to run a limited number of simulation runs varying the design variables in a structured way to generate a set of points on the optimization surface and from these points generate an approximate surface (the response surface) where points and derivatives (sensitivities) are easily calculated and then run the optimization on this surface. More details about optimization can be found in Chapter 5.

1.4 The Structure of the Book

In Chapter 2 we present some aspects of knowledge based engineering such as what KBE is and how to gather knowledge for implementing KBE. Chapter 3 presents simulation theory based on finite element formulations. Chapter 4 present the details of how KBE is implemented for mechanism design. Chapter 5 presents the basics of optimization theory including some implementation aspects for mechanism optimization. Chapter 6 presents the tools that are used to implement the KBE application. Chapter 7 presents the proposed user interface for the KBE application. Chapter 8 presents cases of use of the KBE application and Chapter 9 discusses the findings and gives some conclusions.

2

State-of-the-Art Knowledge-Based Engineering

Many of the findings presented in this chapter have their origin in a La Rocca paper [46]. Other literature that was consulted includes [7, 8, 64, 48, 27, 41, 40, 42, 39, 1].

There are different opinions of what distinguishes a KBE system from a CAD system, but as was mentioned in the introduction to this book, true KBE will require a programming language. The pilot implementations of KBE for this book are implemented in the KBE language AML [60, 22] and more details about this language are presented in Chapter 6. Other KBE languages with similar functionalities are:

- IDL, the ICAD Design Language, is based on a Common Lisp version, with Flavors object system (an early object-oriented extension to Lisp, developed at the MIT Artificial Intelligence Laboratory).

- GDL, the General-purpose Declarative Language of Genworks, is based on the ANSI standard version of Common Lisp and makes use of its CLOS (Common Lisp Object System).

- Intent!, the KBE proprietary language developed by Heide Corporation and now integrated in Knowledge Fusion, the KBE package of Siemens NX (formerly Unigraphics), belongs also to the family of Lisp-inspired languages.

All these KBE languages, including AML, have their origin in Lisp. According to Wikipedia, the Lisp (List Processing) programming language dates back to 1958 and is the second oldest programming language. Lisp became the favorite programming language for artificial intelligence (AI) research. This connection between KBE and AI is not a coincidence. You could argue that all modern software systems are knowledge based. In classical information systems the knowledge is embedded in the software, while in knowledge systems the knowledge is usually defined in an explicit representation in a repository and with utilities that can act on the stored knowledge. Such a repository is often called a knowledge base, and the utilities can be inference engines or reasoners. Knowledge-based systems are a branch of artificial intelligence, and the offspring from AI that have had the highest commercial success. KBE could be regarded as a combination between AI and CAD. All the KBE languages referred to above are within the object oriented (OO) programming

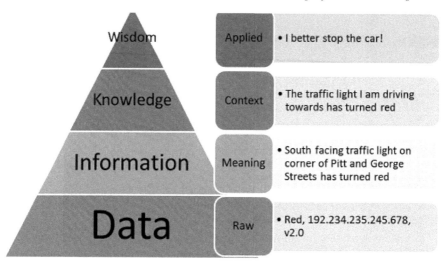

FIGURE 2.1: The knowledge pyramid [1]

language paradigm, but are distinguished from other OO languages like C++ and Java with an extensive library of geometric modeling features for parametric geometry manipulation.

For a more comprehensive state-of-the-art discussion of the KBE technology, refer to La Rocca's paper [46].

2.1   Definitions and Classifications

The concept *knowledge* deserves some discussion in connection with KBE. As mentioned earlier, when designing CAD geometry interactively the knowledge required to do the forming and sizing of the geometry sits in the head of the designer and what this knowledge "looks like" is not well defined. However, when implementing a KBE application this knowledge needs to be made explicit in a program sequence to generate the geometry automatically. We will not try to give a definition of the term knowledge, but we present the knowledge pyramid in Figure 2.1 as an effort to facilitate the understanding of the concept knowledge. This figure demonstrates that the raw data need additional information to give them some meaning, but to use the information for something you also need a context that represents the knowledge level. The top level of the pyramid is Wisdom that says something about what this knowledge could be used for.

Another angle of the concept knowledge is if it is computer-interpretable;

that is if the knowledge is in a form to be applied to problem instances by the computer. A human user usually exercises some element of control over the process, but the knowledge representation is in some sense executable by the computer. The counterpart is human-interpretable knowledge. We confine our discussion to systems involving computer-interpretable representations and CAD software environments. Human-interpretable representations, on the other hand, might interface with the CAD environment. It is unlikely that deeper integration would be pursued as long as the human user is assumed to be in the knowledge processing loop.

There have been several attempts to give a definition of KBE, but here we only present La Rocca's [46] extended definition that includes most of the other attempts for a KBE definition:

Knowledge-based engineering (KBE) is a technology based on the use of dedicated software tools called KBE systems, which are able to capture and systematically reuse product and process engineering knowledge, with the final goal of reducing time and costs of product development by means of the following:

- Automation of repetitive and noncreative design tasks

- Support of multidisciplinary design optimization in all the phases of the design process

Efforts also been made to classify KBE systems according to their type, and the most used classification was given by Penoyer et al [39]:

- *Generative KBE systems* are used to create or synthesize detailed geometry from specifications, rules, predefined constraining geometry and user inputs. Often the geometry is generated as a CAD-like model within what is usually called a KBE framework. As an example: A major automotive company has developed a system that creates the detailed geometry of a hood reinforcement panel based on the hood outer surface geometry and specifications of certain design parameters. Manufacturing and design knowledge (e.g., minimum bend radii, maximum draw depths, etc.) sufficient to automatically generate the detailed reinforcement geometry is utilized.

- *Advisory KBE systems* use design and/or manufacturing knowledge to evaluate designs when they are created, usually by an engineer. As an example: Several systems exist that critique the design of injection moulded plastic parts, assessing the strength, manufacturability and cost of the design, and recommending design changes to improve the characteristics. These systems are capable of recognizing non predefined features in the part geometry, and they apply computer-interpretable knowledge to critique the design for performance and manufacturability.

- *Innovative KBE systems* usually employ model-based reasoning to explore a large design space and locate feasible designs. One well-known system of this type is made for developing automatic transmission gear train designs. Current work is directed at exploring larger gear train design spaces. Multiple candidate solutions are generated and then reviewed by the user.

- *Selection KBE systems* use domain knowledge in combination with user input to assist the user in selecting among many options presented by the KBE application. A KBE selection system might be used for a material selection application, where factors such as durability, weight and other material properties in combination with cost and availability information are used to select the best material from a long list of candidates.

KBE systems can seldom be classified strictly within one of the four classes of KBE systems referred to above, but are often a combination. However, one of the classes referred to is often dominant within a KBE application. For example a generative system might also provide advisory capabilities using the same knowledge base to allow the user to make subsequent modifications to the generated geometry.

2.2   Features of KBE Languages

As mentioned earlier the KBE languages referred to all belong to the object oriented (OO) programming language paradigm. When generating geometric models in a KBE language, for each data object in the actual product generated by the KBE application, there is a one-to-one relation to a physical object. This makes it very easy for the designer, who very often has no programming experience, to see the relations between the data models and the physical product. This makes it possible for a user with no programming skills to operate very large and complex product structures in the KBE application like in a CAD application, to refer for instance to large floaters for offshore oil platforms. An operator of the KBE application should be able to use commands like *create-model* as a high level interface to the programming language that will generate an object in the KBE application representing a physical object. KBE operators like *create-model* are features that are not present in standard OO language and represent an exclusive characteristic of KBE technology. These features concern the ability to generate and manipulate geometry.

Caching refers to the ability of the KBE system to memorize at runtime the results of computed values, for instance computed properties and instantiated objects, so that they can be reused when required, without the need to recompute them again and again, unless necessary. The dependency tracking mechanism serves to keep track of the current validity of the cached values. As soon as these values are no longer valid (stale), they are set to unbound and recomputed if and only at the very moment they are again demanded. As an example refer to the drawing of the mechanism links in the KBE pilot application used for this book. Say center coordinates or direction for a joint is changed and the dependency tracking mechanism is notified by the system that the actual joint has no valid geometry stored. The dependency tracking mechanism is also notified that the link connected to the actual joint has no valid geometry either and to draw the link the actual joint and link geometry needs to be regenerated. For links not connected to the joint that was moved the geometry model is valid and may be drawn directly from the exiting geometric model.

Usually a KBE system has two possible ways of operating, namely by eager or lazy evaluation. This means that when a change is made the system could either update all dependent parameters immediately (eager), like for dependencies in an XL sheet, or postpone (lazy) the updating until the actual value is needed and then update. KBE applications usually use lazy evaluation and that saves computational power by avoiding unnecessary computations; refer to the example of drawing links above. Invalid data or objects are not recalculated unless required for some operations like drawing links.

A KBE application is driven by a set of rules and according to La Rocca [46] the following main rules can be identified:

- *Logic rules (or conditional expressions):* Apart from the basic IF-THEN rule (production rules), KBE languages provide more sophisticated conditional expressions, often inherited from Common Lisp. Example: IF someone knocks on door THEN I will open it.

- *Math rules:* Any kind of mathematical rule is included in this group, including trigonometric functions and operators for matrices and vector algebra. Many others are functions and operators provided by the given KBE language. These rules are commonly used for evaluating computed-properties and child object inputs. Of course, mathematical rules can be used both in the antecedent (before) and consequent (after) part of any production rules.

- *Geometric manipulation rules:* In this category rules for generation and manipulation of geometric entities are included. These are language constructs for the generation of many and diverse geometrical entities, ranging from basic primitives (points, curves, cylinders, etc.) to complex surfaces and solid bodies. It is important to note that all instantiated geometry entities are real objects; as such they can answer messages like length, area, volume, center of mass, etc.

- *Configuration selection rules (or topology rules):* These rules are actually a combination of mathematical and logic rules. However, they have a different effect than just evaluating a single numerical or Boolean value, and thereby they deserve a special label. They are used to change and control dynamically the number and type of objects in an object tree. Hence they can affect the topology of any product and process KBE model.

- *Communication rules:* In this group all the specific rules that allow a KBE application to communicate and/or interact with other software applications (not necessarily KBE) and data repositories are included. Rules exist that allow accessing databases or various kinds of files to parse and retrieve data and information to be processed within the KBE application.

Other rules exist to create files containing data and information generated by the KBE application. For example, it is possible for a KBE application to generate as output standard geometry exchange data files like STL, IGES and STEP; or XML files or any sort of free format ASCII files for more textual data. Rules also exist to start at runtime external applications, wait for results, collect them and return to the main thread.

Handling and manipulation of geometry is another very powerful feature of KBE systems that is not available in any other type of knowledge system.

2.3 Knowledge Acquisition for KBE

There are several ways of structuring KBE systems and different approaches must be selected in different cases. Even so, there are still some general criteria that should be fulfilled in any case:

- Knowledge has to be present

- This knowledge has to be acquired (knowledge acquisition)

- It needs to be structured into a knowledge database

- You need to establish some kind of common library (with standardized parts) based on the knowledge that is acquired

In the following we will take a look at how to construct a KBE system. Since most of the literature on this subject is rather discipline-specific, the approach is taken from the ship industry, but it still is pretty general and transferable to other disciplines. Knowledge is a vital component of engineering design. The contents of a knowledge base can be used in a number of ways:

- To disseminate knowledge to other people in an organization

- To reuse knowledge in different ways for different purposes

- To use knowledge to develop intelligent systems that can perform complex design tasks

Knowledge is hard to acquire from specialists. The difficulties stem from a number of factors since specialists:

- are not good at recalling and explaining everything they know

- possess what is termed tacit knowledge which operates as second nature, usually hard to explain without acting upon it

- have different experiences and opinions.

Also, specialists:

- develop particular conceptualizations and mental shortcuts that are not easy to communicate

- use jargon and assume most other people understand the terminology they use.

In addition to these difficulties are other problems associated with representing and storing knowledge, such as the granular size of the knowledge and the representation formalism. It is a key element of knowledge-based engineering that knowledge is identified and represented in different forms making up the knowledge base. As mentioned in the introduction to this chapter, knowledge engineering is a kind of artificial intelligence (AI) technology, which uses the principles and methods of AI for providing a means for solving difficult problems. The core of knowledge-based engineering is to integrate:

- professional knowledge

- domain knowledge

- user's maturity design experience

- the choice of design parameters based on experimental data

- material data

- user's feedback

- relevant design standards and norms into the design of software through logical judgments and deduction

to achieving product intelligent design.

Moreover, knowledge base management realizes the accumulation and update, avoiding loss and outdated knowledge. Summing up knowledge-based engineering includes:

- knowledge acquisition

- knowledge representation

- knowledge reasoning

with its emphasis on reusing product design knowledge, experience and other kinds of knowledge in the design, developing new and optimal products at top speed.

According to Milton [30] knowledge acquisition for KBE consists of three parts:

- *Knowledge Capture* is concerned with identifying and eliciting knowledge that might be included in the knowledge base. The source of knowledge can be human experts, written material, images, models source code, video or any other useful knowledge source. Capturing knowledge from human sources is referred to as knowledge elicitation.

- *Knowledge Analysis* is performed by the knowledge engineer after capture sessions. The captured knowledge is broken down into elements to be put in the knowledge base to create a structure suitable to the given domain. The elements are referred to as "knowledge objects", and can be, for example, concepts, values, attributes or relations. Analysis also includes selecting which elements are necessary to implement in the knowledge base.

- *Knowledge Modelling* is creating different ways of viewing the knowledge in the knowledge base. Each view gives a certain perspective on the knowledge. For instance the assembly of parts in a pen can be viewed in a hierarchy, while the connections between objects and attributes, for instance which parts are driven by the attribute "diameter" and which are driven by the attribute "spring tension" can be viewed in a matrix.

Knowledge acquisition refers to acquiring knowledge access to resolve problems encountered in the development from knowledge sources (such as design standards, product specifications, expert knowledge, experience and successful precedents (earlier events), etc.) in the domain. This knowledge mainly exists and is generatively imported in the form of kinds of formulas, laws, design tables, rules, checks, reactions, loops, generative scripts and so on. Knowledge representation researches how to state problems and how to store knowledge in a machine-interpretable representation to facilitate computers to take advantage of knowledge in the knowledge base to address complicated and difficult problems.

Knowledge-based reasoning is the thinking process which is deducing another judgment from judgment known according to a strategy. Designers express the product knowledge base including expert knowledge, experience knowledge, specifications and precedent through knowledge-based engineering to guide the designer in the design process. The reasoning methods include rule-based reasoning (RBR) and case-based reasoning (CBR). The method of RBR is mainly used for specific parameters based on a knowledge adviser and knowledge expert work processes. The method of CBR is used for product and part design similar to original case-based product knowledge templates.

Besides, the friendly interface provided by the software system helps users export and import design information and manufacturing information to achieve a joint deliberation of human–computer to address problems.

Using KBE can significantly improve the work-flow of the design. Designers can learn the rules and regulations from the knowledge base in the modeling stage. The knowledge base system can provide not only proper references, suggestions and support but also integrates the design, especially the rules related to the design which can avoid lots of the mistakes in design and reduce the preproduction time. By increasing the quality of the design and modeling, the whole building time can be reduced dramatically. The knowledge base is defined as a collection of experience, rules, cases and other knowledge. In knowledge-based engineering design methods, knowledge including:

- expert knowledge,

- experience and product design standards,

- collecting product specifications and successful precedents,

- sorting and summarizing into a number of rules, analysis and problem-solving strategies,

is placed in a particular form of the document or database to constitute a knowledge base, which can achieve storage and classification management of product design knowledge and provide best guidance and recommendations for the designer during the design process.

Significant improvement in efficiency and quality can be realized, if the knowledge base can be used in a number of ways:

- To communicate knowledge with fellow workers and disseminate knowledge to other people,

- To reuse knowledge in different ways for different purposes,

- To use knowledge to develop intelligent expert systems that can perform complex design tasks.

Moreover, taking advantage of a variety of established knowledge about the product design and process, the workload of the designer is eased, instead of determining rational solutions for similar problems over and over again, so the conventional design process can be accelerated by using knowledge-based engineering.

A ship as a product belongs to a large comprehensive product list of various product attribute variations, with a complex structure, great amounts of outfit items and size information. The technology in the various shipyards includes production capabilities and product varieties of ships. A library with standard parameterized components is part of the knowledge base for the ship design process. New structural parts can be adjusted in parameters and designers can update the standard parts library whenever necessary. Standard parts characteristic parameters are stored in Excel tables and other relevant information about the component is saved synchronously for designers to exchange and share. So by entering appropriate values to adjust standard features dimension-driven or feature-driven goals can be achieved. The relationships between standard parts are established through publication tools in the form of design tables, formulas, checks and rules, and equations. The standard parts are saved in the form of document templates. In the process of application of CBR for standard parts design, it needs to choose a case with a similar design goal, select reference geometry and change parameters.

2.3.1   Tools for knowledge acquisition and modeling

In the following sections methodologies for knowledge acquisition for KBE systems are introduced. Two of the methodologies, CommonKADS and MOKA, also include different approaches to the entire KBS development process, and an overview of these processes will be given. The main focus will be on how to handle knowledge acquisition and modeling.

CommonKADS: Knowledge Acquisition and Documentation Structuring [14]. CommonKADS is not restricted to KBE or engineering domains, but aims to be a general methodology for knowledge engineering with benefits for all types of knowledge-based systems (KBS) as well as knowledge management activities. Much of the work done on specialized methodologies for knowledge systems is based on ideas from CommonKADS. KBS does not operate in an organizational vacuum; therefore, it is necessary to keep track of the organizational environment which the system operates in to develop and utilize it successfully. As a consequence CommonKADS not only defines models and tools for analysis and modeling of a domain for the KBS, but also includes descriptions on how to model the organization, business processes and task executors.

MOKA: Methodology and tools Oriented to Knowledge-based engineering Applications [58]. Its main goal was to develop a methodology especially suited for KBE to reduce the amount of upfront investment and risk associated with the development of KBE applications. The MOKA initiative defined an application development cycle, called the KBE life-cycle; see Figure 2.2.

The development begins with the identification phase of a problem or need. The problem can for example be a bottleneck in the development process. Feasibility studies are performed, including which tools and techniques might be relevant to solve the problem. The justification phase deals with risk assessment, costs, project planning and to define criteria to evaluate the success or failure of the project. The capture phase is when necessary knowledge is collected, analyzed and modeled. The formalize phase involves creating models for implementation, e.g., UML-like diagrams, pseudo code etc. The package phase is concerned with coding the system on the chosen platform, and finally the activate phase puts the system in practical use in an organization. The

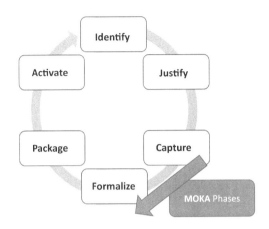

FIGURE 2.2: MOKA KBE life-cycle [58]

MOKA cycle continues as new opportunities for the application are identified requiring maintenance or development. The MOKA phases in the KBE
life-cycle are the capture and formalize phases. MOKA briefly describes the
rest of the steps, but does not go into any depth in any of them. To facilitate
knowledge models both on the knowledge level and a representation on the
symbol level, two "layers of representation" are used in MOKA. The layers
are called the informal and the formal models. The basic idea is first to create
knowledge level models in interaction with experts, then to translate these
to formal models for implementation. The informal and formal models are
described briefly in the following.

The informal model is made up of four parts, each contributing to the complex description of the domain knowledge independent of any implementation
details. The four parts are:

- ICARE forms,

- Charts,

- Terminology (a glossary),

- A knowledge book.

The ICARE (Illustrations, Constraints, Activities, Rules and Entities) forms
are predefined templates that enforce classification of knowledge into 5 different categories. Charts create a visual representation of the knowledge types

and connection between them. The glossary is defined simply to help rule out misinterpretations and disambiguations. The Knowledge Book is a document compiled from the forms, charts, glossary and supplementary comments.

The formal model is based on the informal model. A formal model is constructed using UML-like notation called MML, for Moka Modeling Language. The goal is not so much to create a coder-friendly representation, as it is to prepare for a machine-readable and checkable representation. The main elements of the formal model are the product model and design process model. The former is split in five views, constructed as UML class diagrams and the latter is based on activity diagrams.

During the development of MOKA, it was tested on small-scale engineering processes in the auto and aero industries. According to the developers of the methodology/tool there should be no reason why MOKA should not work on large application domains. But in the final MOKA project review, industrial partners pointed out that the methodology as it is does not support large scale-applications.

Currently, neither the MOKA methodology nor the MOKA tool is powerful enough to manage larger knowledge bases. This is not only a quantitative problem. Engineering knowledge is typically characterized by many interdependencies and cross relations, which have to be treated carefully. This results in high requirements to knowledge modeling in general and especially to validation, verification and modularization [11]. Besides, no industrial application of MOKA has yet been reported.

Elicitation aims to capture knowledge from experts, and make it explicit and stored for further analysis, structuring and coding in a knowledge base. This is done through some sort of direct interaction with the expert. The difference between knowledge elicitation and knowledge acquisition is that while acquisition refers to gathering knowledge from any source, elicitation is the subtask of acquisition from human experts. Elicitation techniques can be split into natural and contrived. Natural methods are considered familiar to the expert, a method she would normally use by herself to describe knowledge. Contrived are the methods the expert is not used to participating in when exhibiting knowledge, and would normally not use to present knowledge. As a natural technique, interviews are the techniques used most often, and are split into structured, unstructured and semi-structured types, according to the degree of freedom in the communication between the elicitor (knowledge engineer) and interviewee (expert). Protocol analysis (also called commentary) is a term for a number of different ways of analyzing the expert while actually solving problems in the domain. The expert is observed while doing a task, and comments on everything while being observed (on-line) or her actions are recorded and she can comment on them afterwards (off-line). Focused discussion is similar to interviews or meetings. An item, which can be an object, model, graph, a document, etc., is used as a point of focus in the discussion. Teach back is a method for validation, and means that the elicitor tries to teach the knowledge back to the expert or to a person not yet introduced to the subject. Critical decision method (CDM) is used to elicit knowledge of a task that an expert has performed numerous times before. The technique focuses on eliciting the basis for making perceptual and conceptual discriminations, typicality judgments and critical cues.

Regarding contrived techniques we refer to methods such as concept mapping that are ways of discovering how the expert sees relations between concepts in a domain. The methods are often performed by sorting or grouping cards with concept names printed on. Laddering techniques are another somewhat contrived way of obtaining classifications. The elicitor and expert create a hierarchical breakdown/tree of the domain at hand to classify a domain and reveal new concepts. Grid techniques are used to record and discover relationships between concepts, attributes, items, objects etc. The items at hand are set up as rows and columns, and the relationships are recorded in the cells. Twenty questions (also called limited information task) are used to discover the expert's strategies in problem solving. It is performed the same way as the parlor game, with the expert asking questions of the elicitor to reveal what the elicitor is thinking about. Triadic elicitation is a way of revealing attributes through describing similarities and differences between items in a domain. Triadic means that three items are compared. New attributes are discovered through describing how two of the items are similar and how the third is different from the other two.

Recording elicitation sessions: Several authors on knowledge elicitation recommend audio or video recording of elicitation sessions [30, 49]. Audio recordings have several benefits. They serve as a perfect record of what is said; nothing is missed. They reduce the need to take notes (which can contain errors and miss vital knowledge) and they serve as a valuable knowledge source for analysis and modeling. The recordings can be transcribed afterwards. But transcription is a time-consuming process, and complete transcripts are seldom possible due to the time limitations on most projects. Partial transcripts are more often feasible, and Milton [30] points out that just looking through the transcript once refreshes the memory of the knowledge engineer.

Choosing the right techniques: As a general principle, the knowledge engineer will start out with interviews and choose other methods when considered useful. It is generally normal to work from common knowledge towards more specific problems. It is difficult to find well-documented information on which technique to choose when, and the order the different techniques should be used in. Milton's [30] advice is to prepare for several techniques in one session, and balance the need to unlock knowledge and keep the expert happy:

Each technique is like a key. You sometimes have to try a few keys in the lock before one turns it and allows the knowledge to flood out. Experience will teach you which keys to take with you and when to switch from one to another. [30]

The rule interchange format (RIF) designed in LinkedDesign [26, 23] has three levels:

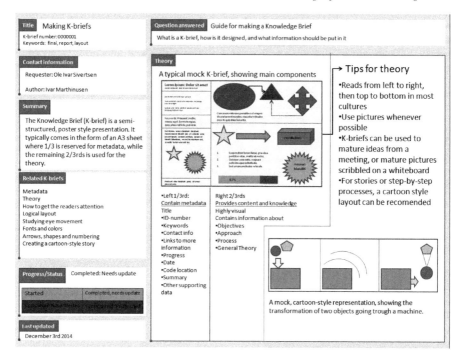

FIGURE 2.3: Knowledge brief [29]

- The K-briefs layer—the A3 format knowledge briefs (refer also to the informal level in MOKA discussed above),

- The K-model layer—the extended SysML modeling (refer also to the formal level in MOKA, the MOKA Modeling Language (MML)) and

- The K-code layer—KBE implementation (Ex. AML [60] source code).

This approach could be referred to as an alternative to CommonCADS or MOKA. The K-Brief template should be used to capture the knowledge in a visual and interactive way, but without losing context- or meta-information. The K-Brief template itself has been further elaborated by the LinkedDesign [26, 23] consortium. The structure and concept of the proposed K-Brief template is drafted in Figure 2.3.

The established standard SysML [37] is proposed for the K-Model layer. The region marked Extensions to SysML (RIF Profile) in Figure 2.4 indicates the new modeling constructs defined for RIF that have no counterparts in SysML (or UML).

The third layer of RIF represents the implementation of the KBE system in a KBE programming language, e.g., the AML framework. Here the codified knowledge can remain in a proprietary format (e.g., LISP syntax as used in AML [60]). If the code itself is based upon object-oriented programming,

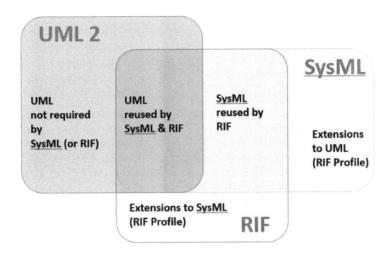

FIGURE 2.4: Rule interchange format [43]

programming code snippets can directly be linked to elements of the K-Model layer. It can be argued that the RIF format will have many of the same drawbacks as the MOKA method (refer to the MOKA critique [11] referred to above): Could work on small applications, but could be hard to apply on complex systems. The question is should we look for an alternative to the proposed methodologies: CommonCADS, MOKA and RIF?

Another alternative for knowledge acquisition that is considered is the OPM (Object Process Methodology) language [13] for system modeling that has lately become an ISO standard. The OPM language is very easy to understand and use and is now considered as a possible alternative for knowledge documentation for knowledge acquisition (KA).

The experienced people at the KBE Company TechnoSoft Inc. have a somewhat different approach. Their experience is that KBE applications are so different that a structured approach is very difficult. Some of the ideas from discussions at TechnoSoft are presented in the following. Working in teams developing actual KBE applications a few generic principles have been established:

- When a project has been established for developing a KBE application the domain expert should just produce all the information about data and processes that is available and present it to the KBE expert. It is seldom useful that the domain expert tries to edit the data to a form that is more fitted for the KBE implementation. The KBE expert should study the data that is available and based on that specify what more is needed. In this way the time spent by the domain expert is more efficient and effort could only be spent on issues of importance for the KBE implementation.

- The KBE expert will very often like to start to study the documents that are produced in the backend of the manual process to be automated. For large engineering projects, refer for instance to offshore, the work is often compartmentalized and starting from the final deliverables from the project, the KBE expert will be able to see the big picture and disclose the pieces of information that are not explicit in the documentation that is available.

- In an early stage the KBE expert will start to look for bits and pieces from earlier applications and try to put together pilot application modules that could fit into the final KBE application. In this way the effort necessary to develop the new application could be reduced by building on tested modules from earlier applications.

- It is seldom fruitful to ask the domain expert what automated results he would like. The KBE expert should produce the results he thinks are right from the big picture he sees and the domain expert should give feedback on what is wrong. This is the most effective use of the domain expert's time.

- Establish good relations between the KBE experts and the domain experts and appoint key persons on both teams.

- Making a new KBE application is an extensive learning process not only for the KBE experts, but also for the domain experts. The domain experts will often be challenged to produce information that is implicit for the manual engineering process, but is crucial to make explicit when automating the processes.

- We do not believe that separate tools and formats for knowledge acquisition and modeling are of much use. This only complicates the route between the domain expert and the final KBE application. The shorter and simpler the route between the origin of the knowledge and the KBE application that produce the results, the better.

- Collaboration tools and methods for communicating the knowledge could be of great value, especially because you often are dependent on electronic communication. Uses of A3 reports, also called knowledge briefs (K-Briefs), for documenting the actual acquisition process are being considered [29]; see Figure 2.3.

An obstacle against introducing a tool for knowledge acquisition is that KBE systems are developed by programmers and there is often a strong opposition against introducing rigid documentation procedures besides the source code.

2.4 Developing KBE Applications

As described in the previous section, gathering the knowledge from experienced engineers (domain experts) is the main challenge when developing a KBE application. How to do this is not straightforward and, as discussed in the previous section, no preferred method could be recommended. The best approached is probably to follow the principles used by experienced developers; see previous section. We will limit our discussion here to the task of developing a KBE application using the AML framework from TechnoSoft [60].

When you have acquired a certain level of knowledge from the domain experts you should look for existing modules that could be used in the planned application, modules that could be used directly or modules that could be extended to fit in the new application. A module could be a standalone KBE application or just a cluster of KBE classes that could be used as is or updated for the new application. Reuse is an important part of KBE and reuse of existing modules or cluster of classes is an important part of this. One important advantage of this is that you can use code that is already tested in another application and another advantage is that you could rapidly put together a pilot application to produce some results for the domain expert. The best way to prove for the domain expert that you did understand the knowledge correctly is by producing output. Because some of the knowledge you acquired from the domain expert usually could be named "tacit" knowledge that is difficult for the expert to explain and make explicit. This is information that you very often misinterpreted in the early stage, but will be easier for the domain expert to disclose when he sees the errors in the early pilot implementations. The loop of piloting, producing results for the domain expert that has errors, correcting the pilot for producing improved results and so on is a well-tested way of working when developing a new KBE application. For this reason experienced developers of KBE applications are very important.

For a new company that considers developing KBE applications it could be advantageous, or even necessary, to work together with experienced KBE developers as consultancies making their first KBE applications. This is also important for getting an application running as soon as possible and having a rapid feedback within the company regarding benefits. If the company plans to develop more KBE applications in the future it is recommended to be involved in all phases of the KBE development, not only the knowledge acquisition activity but also the implementation activity to build up internal knowledge to be reused in future applications.

There are many pitfalls when starting to introduce KBE in a company. The designers will not necessarily see this as a good idea. First of all they may see this as a threat to their jobs. Automating routine design work may be seen as a way to lay off people, especially in times when the company is struggling with the economy. Arguments that could be useful are that KBE will be very beneficial for product quality and performance because more alternatives could be investigated for less time and cost. This could give a competitive edge in the market, besides, not using KBE to improve the company's performance could jeopardize the company's existence. The alternative of moving activity to a low-cost country could also be mentioned as an argument for KBE. Without having the experienced engineers (domain experts) fully committed to KBE implementation, it could be very difficult to make this happened. Another important argument is that introducing KBE could move more resources into creative work and in this way also strengthen the company.

An experience worth mentioning is that the people gathering the knowledge for the KBE application (knowledge experts) should not be pure programmers without good knowledge of engineering terms. This would make the engineers very reluctant to work with these people because of a very difficult communication process. If the knowledge expert is both skilled in engineering and in computer science, the knowledge acquisition will be very smooth. For an efficient process there should be a good team spirit between the engineering experts and the knowledge experts.

Typical for implementing a KBE application is to focus on the algorithms within the application to make the "KBE engine" work correctly. The user interface is often made as the last step of the implementation; that is quite unusual in software development in general. This is probably the reason why KBE applications are seldom very "sexy" from the user's side; however, this could be different if the user interface is given higher priority. The reason for the user interface being implemented last is that the KBE engine inside could require new work processes for engineers. Because of the automation in a KBE system the detailing of the design comes very early; i.e., the preliminary steps of design are integrated with the detailing.

2.5   Knowledge Acquisition for Introducing KBE in Mechanism Design

The content of this section is to a large extent based on the work of Thor Christian Coward [10] for his master's thesis.

As discussed in Section 2.3.1, developing KBE applications is often so different from case to case that a very structured approach is seldom beneficial. Introducing automation (KBE) into mechanism design is an example of a case that we think is quite different from other KBE applications. First of all a KBE application is usually developed for one company and solely used within that company. The KBE application developed for automation in mechanism design is not made by a certain company for use in the same company, but a more generic application for mechanism design that has a potential to be used by many companies working with mechanism design. The AMETank[1] KBE application developed by the TechnoSoft Company [22] has some resemblance in the sense that this KBE application is used by many companies, however, the field of application is quite different.

The goal for the KBE application developed for our project is to automate major parts of the generic mechanism design process, especially when dynamic simulation is used as a tool both for verifying the system's functional requirements and the structural integrity of the same system. Gathering knowledge about mechanism design in general is a quite straightforward process because this is probably the oldest field of engineering that has been researched for hundreds of years. As will be discussed into more detail in Chapter 4 a mechanism is defined by a set of links connected to each-other with joints that control the relative motion between the links.

Dealing with the kinematics of mechanisms, links are assumed to be rigid, and thus a link is defined as *the rigid connection between two or more elements of different kinematic pairs* [62]. Links can be in a wide range of shapes and forms, but, the connection points are always relatively fixed to each other. The links are connected to each other by kinematic pairs, also called joints, constraining relative motion. And, links are classified by their number of joint connections, referred to as the links degree.

Joints sharing two mating surfaces, one for each link, therefore are said to form kinematic pairs. Reuleaux [45] divided the pairs into lower pairs and higher pairs. Lower pairs share surface contact, while higher pairs only share a line or point contact. There are only six pairs classified as lower pairs: revolute, prismatic, helical, cylindric, sphere and flat. The lower pairs are illustrated in Figure 2.5 and further presented in Table 2.1 with their pair symbols (as proposed by Denavit and Hartenberg [12]), pair variable(s) and degrees of freedom. While there are only six lower pairs, there could be an infinite number of higher pairs, e.g., a ball rolling on flat surface, mating gear teeth or a belt and pulley. As may be seen in Figure 2.5 for the lower pairs each joint has two coordinate systems attached, one xyz coordinate system for the ingoing link and one uvw coordinate system for the outgoing link. The relation between these coordinate systems in each joint is a transformation matrix as a function of the joint variables. Joint variables are rotational and/or translational variables controlling relative motion in the joint; see also Table 2.1 and the Sheth–Uicker (SU) formulation [63].

The topology of a mechanism is how the mechanism links are connected by joints. Each joint is connecting exactly two links is a general observation if we also define the foundation (ground) for the mechanism as a separate link.

As stated in the introduction to this book the main goal for this work is to develop a KBE application to automate the generation of input data for a simulation tool for dynamic simulation of mechanisms both for verification of the mechanism design requirements and to verify the structural integrity

---

[1] AMETank: http://www.ametank.com/

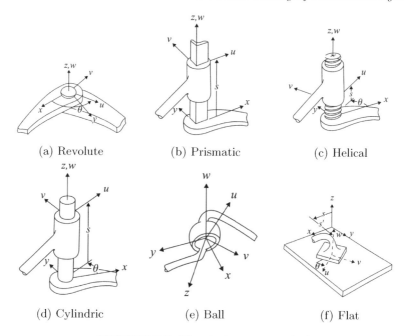

(a) Revolute   (b) Prismatic   (c) Helical

(d) Cylindric   (e) Ball   (f) Flat

FIGURE 2.5: The six lower pairs [63]

of the system [3, 51]; see also Chapter 3. The classical mechanism theory presented above is the basis for the development of the KBE application, but what is typical for any KBE development is that you will need to make some decision how generic your KBE application should be.

All the lower pairs are candidates for implementation in our KBE application. Of higher pairs the gear transmission and the cable and pully connection are candidates for implementation. What is prioritized first for implementation is what gives most application potential. The KBE application needs algorithms to generate geometric models for every joint. A joint will have two parts sitting on different links as for the revolute joint with the axle connected

TABLE 2.1
The lower pairs, presented with name, symbolic notation, mathematical variable, degrees of freedom and relative motion.

Pair	Symbol	Pair Variable	DOF	Relative Motion
Revolute	R	$\Delta\theta$	1	Circular
Prismatic	P	Δs	1	Rectilinear
Screw	S	$\Delta\theta \, or \, \Delta s$	1	Helical
Cylinder	C	$\Delta\theta \, and \, \Delta s$	2	Cylindric
Sphere	G	$\Delta u, \Delta v, \Delta w$	3	Spheric
Flat	F	$\Delta\theta, \Delta x, \Delta y$	3	Planar

to one link and the hub connected to another link. The axle and the hub connected defines the revolute joint, also named the male and the female parts of the joint, respectively. Coming up with generic algorithms for generating geometric models for joints is quite straightforward depending on how much detail should be included. The simulation tool referred to above also has some additional joints not referred to in classical mechanism theory as the free joint and the rigid joint. The free joint has no constraints and the Rigid joint has no degrees of freedom since it has 6 constraints and no relative movement possible in the joint. The implementation of these two joints in the KBE application should also be given high priority.

As discussed above, classical joints (lower pairs) are well defined and algorithms for generating geometry is just a matter of effort. For mechanism links it is a quite different story. What we know about a link is that it is connected to a number of joints positioned in space and for each joint the link will host either a male or a female joint part; see definition above. For a link to be part of the mechanism motion it will need to be connected to two or more joints. The challenge that needs to be solved is to be able to generate mechanisms automatically based on some defined points in space (e.g., joint positions), i.e., to generate the geometry of each link based on a limited set of parameters. Details on how this is defined and implemented are presented in Chapter 4. In the following is a short description of the strategy used. A link can be connected to 2, 3, 4 or more joints and the strategy is to define connections between the joints in the link, called members. For 2 joints in the link only one member could be defined with a cross-section and a variation along the member. The member may be extruded along a straight line between the joints or along a spline curve given by some weighted points in space. If the link is connected to 3 joints, three members are possible, but the link could also be defined with only two members, i.e., one of the members could be suppressed. The members chosen will be defined as for the binary link above (the link with 2 joints). For a link with 3 joints you could keep all possible members, but for a link with 4 or more joints you usually suppress a number of potential members. If the members form a loop as for the link with three joints, a surface will also be generated between the members.

For each link, the link's joint parts and the links members are glued together to one solid part using the *union* operation of the geometric modeler. With all links modeled this way with springs and dampers modeled in joints or between links, including possible loading, the geometry is ready for generation of simulation input. The input includes blending of sharp edges and FE meshes for each link and connectivity data for the mechanism.

For the design loop to be fully automated, control engineering including function definitions input should also be modeled for the KBE application.

As mentioned above, the details for the geometric modeling in this KBE application may be found in Chapter 4.

3

State-of-the-Art FE-Based Dynamic Simulation

3.1 Description of Motion

3.1.1 Notation

A vector (first-order tensor) $\boldsymbol{a}$ can be expressed in two different coordinate systems as

$$\boldsymbol{a} = a_1 \boldsymbol{I}_1 + a_2 \boldsymbol{I}_2 + a_3 \boldsymbol{I}_3 = \sum_{i=1}^{3} a_i \boldsymbol{I}_i = a_i \boldsymbol{I}_i \tag{3.1}$$

$$= \tilde{a}_1 \boldsymbol{i}_1 + \tilde{a}_2 \boldsymbol{i}_2 + \tilde{a}_3 \boldsymbol{i}_3 = \sum_{i=1}^{3} \tilde{a}_i \boldsymbol{i}_i = \tilde{a}_i \boldsymbol{i}_i \tag{3.2}$$

One has adopted the notation of using italicized bold for tensors, such as $\boldsymbol{a} = \tilde{a}_i \boldsymbol{i}_i$. Upright bold is reserved for matrices and matrix vectors (one column matrices). Using this definition, the vector of (3.2) will have its matrix representations as

$$\mathbf{a} = \begin{bmatrix} a_1 \\ a_2 \\ a_3 \end{bmatrix} \quad \text{and} \quad \tilde{\mathbf{a}} = \begin{bmatrix} \tilde{a}_1 \\ \tilde{a}_2 \\ \tilde{a}_3 \end{bmatrix} \tag{3.3}$$

in the two coordinate systems

3.1.2 Rigid body motion

The position and orientation in space of a rigid body, S, is unique if one has defined 1) the position of a point, and 2) the orientation of a fixed coordinate system.

The vector $\mathbf{s}$ represents the position of the fixed coordinate system in relation to the global coordinate system. The fixed coordinate system is defined by the three orthonormal base vectors $(\boldsymbol{i}_1, \boldsymbol{i}_2, \boldsymbol{i}_3)$, which comprise the rotation matrix

Position:
$$\mathbf{s} = \begin{bmatrix} s_x \\ s_y \\ s_z \end{bmatrix} \tag{3.4}$$

Orientation:
$$\mathbf{R}_S = \begin{bmatrix} \mathbf{i}_1 & \mathbf{i}_2 & \mathbf{i}_3 \end{bmatrix} = \begin{bmatrix} i_{1x} & i_{2x} & i_{3x} \\ i_{1y} & i_{2y} & i_{3y} \\ i_{1z} & i_{2z} & i_{3z} \end{bmatrix} \tag{3.5}$$

It is frequently necessary to describe the motion within the rigid body S, in, for example, joint attachment points. Joint attachment points have a position relative to the rigid body coordinate system. The joint has an orientation relative to the rigid body's orientation. Defining the point N (a node, for example) within the rigid body with coordinates $\tilde{\mathbf{a}}$ relative to the body's fixed coordinate system, a local coordinate system, $\tilde{\mathbf{R}}_N$, relative to the body fixed system is represented by

Position:
$$\tilde{\mathbf{a}} = \begin{bmatrix} \tilde{a}_x \\ \tilde{a}_y \\ \tilde{a}_z \end{bmatrix} \tag{3.6}$$

Orientation:
$$\tilde{\mathbf{R}}_N = \begin{bmatrix} \tilde{\mathbf{i}}_1 & \tilde{\mathbf{i}}_2 & \tilde{\mathbf{i}}_3 \end{bmatrix} = \begin{bmatrix} \tilde{i}_{1x} & \tilde{i}_{2x} & \tilde{i}_{3x} \\ \tilde{i}_{1y} & \tilde{i}_{2y} & \tilde{i}_{3y} \\ \tilde{i}_{1z} & \tilde{i}_{2z} & \tilde{i}_{3z} \end{bmatrix} \tag{3.7}$$

The point N has a motion in the global system produced as a result of the motion of the body's fixed coordinate system. Using $\mathbf{a}$ for the coordinates of the point relative to the global system yields

$$\mathbf{a} = \mathbf{R}_S \tilde{\mathbf{a}} + \mathbf{s} \tag{3.8}$$

and the orientation of the point in the local coordinate system $\tilde{\mathbf{R}}_N$, will be as follows in the global system:

$$\mathbf{R}_N = \mathbf{R}_S \tilde{\mathbf{R}}_N \tag{3.9}$$

These equations can be condensed by using 4×4 transformation matrices containing the position and rotation as defined in (3.4) and (3.5)

$$\mathbf{P}_S = \begin{bmatrix} \mathbf{R}_S & \mathbf{s} \\ \mathbf{0} & 1 \end{bmatrix} = \begin{bmatrix} i_{1x} & i_{2x} & i_{3x} & s_x \\ i_{1y} & i_{2y} & i_{3y} & s_y \\ i_{1z} & i_{2z} & i_{3z} & s_z \\ 0 & 0 & 0 & 1 \end{bmatrix} \tag{3.10}$$

and

$$\tilde{\mathbf{P}}_N = \begin{bmatrix} \tilde{\mathbf{R}}_N & \tilde{\mathbf{a}} \\ \mathbf{0} & 1 \end{bmatrix} = \begin{bmatrix} \tilde{i}_{1x} & \tilde{i}_{2x} & \tilde{i}_{3x} & \tilde{a}_x \\ \tilde{i}_{1y} & \tilde{i}_{2y} & \tilde{i}_{3y} & \tilde{a}_y \\ \tilde{i}_{1z} & \tilde{i}_{2z} & \tilde{i}_{3z} & \tilde{a}_z \\ 0 & 0 & 0 & 1 \end{bmatrix} \tag{3.11}$$

The position and orientation of point a in the global system can then be expressed as[1]

$$\mathbf{P}_N = \mathbf{P}_S \tilde{\mathbf{P}}_N = \begin{bmatrix} \mathbf{R}_N & \mathbf{a} \\ \mathbf{0} & 1 \end{bmatrix} \tag{3.12}$$

3.1.2.1 Rotation

Any rotation of a vector from its original direction $\mathbf{a}$, to its rotated direction $\mathbf{a}'$, can be described by the general relationship

$$\mathbf{a}' = \mathbf{R}\mathbf{a} \tag{3.13}$$

where the matrix $\mathbf{R}$ is a 3×3 orthonormal matrix. A matrix is orthonormal when each row and column representing a vector is orthogonal to all other rows and columns, and has a unit length. Since $\mathbf{R}$ is orthonormal, the inverse relationship becomes

$$\mathbf{a} = \mathbf{R}^{-1}\mathbf{a}' = \mathbf{R}^T\mathbf{a}' \tag{3.14}$$

Rotation matrix $\mathbf{R}$ can be established in a number of ways, known as parameterizations of the rotation. The rotation matrix is also often called the rotation tensor as it has properties of a second-order tensor.

3.1.2.2 Rodrigues parameterization of rotations

Euler's rotation theorem states that in three dimensions, any motion of a body where a single point is fixed, is equal to a single rotation about a fixed axis through this point.

A consequence of this theorem is that any combination of rotation can also be represented as a single resulting rotation about one axis.

Rodrigues parameterization expresses directly the premise of Euler's theorem; The single rotation θ about a rotational axis defined by a unit vector $\mathbf{n}$. With the vector being defined by

$$\boldsymbol{\theta} = \theta\mathbf{n}, \tag{3.15}$$

the rotation matrix can be written

$$\mathbf{R} = \mathbf{R}(\boldsymbol{\theta}) \quad = \mathbf{I} + \frac{\sin\theta}{\theta}\widehat{\boldsymbol{\theta}} + \frac{1}{2}\left(\frac{\sin\frac{\theta}{2}}{\frac{\theta}{2}}\right)^2\widehat{\boldsymbol{\theta}}^2 \tag{3.16}$$

$$= \mathbf{R}(\theta, \mathbf{n}) = \mathbf{I} + \sin\theta\widehat{\mathbf{n}} + (1 - \cos\theta)\widehat{\mathbf{n}}^2 \tag{3.17}$$

where the wide hat notation is defined by

$$\widehat{\boldsymbol{\theta}} = \begin{bmatrix} 0 & -\theta_z & \theta_y \\ \theta_z & 0 & -\theta_x \\ -\theta_y & \theta_x & 0 \end{bmatrix} \quad \text{where} \quad \boldsymbol{\theta} = \begin{bmatrix} \theta_x \\ \theta_y \\ \theta_z \end{bmatrix} \tag{3.18}$$

---

[1]For computer implementation, the 4×4 matrices are represented by 3×4 matrices obtained by deleting the last line of the 4×4 matrices as in, for example, (3.10). The matrix multiplication of (3.12) is modified by taking into account the implicitly defined [0 0 0 1] last row of the matrix.

3.1.3 Variation of Rodrigues parameterization

The variation of the instantaneous rotation axis $\boldsymbol{\omega}$ with respect to the finite rotations $\boldsymbol{\theta}$ of the Rodrigues parameterization can be written as

$$\delta\omega = \frac{\partial\omega}{\partial\boldsymbol{\theta}}\delta\boldsymbol{\theta} = \mathbf{H}(\boldsymbol{\theta})\delta\boldsymbol{\theta} \tag{3.19}$$

where

$$\mathbf{H}(\boldsymbol{\theta}) = \frac{1}{\theta^2}\boldsymbol{\theta}\boldsymbol{\theta}^T + \frac{\sin\theta}{\theta}(\mathbf{I} - \frac{1}{\theta^2}\boldsymbol{\theta}\boldsymbol{\theta}^T) + \frac{1-\cos\theta}{\theta^2}\widehat{\boldsymbol{\theta}} \tag{3.20}$$

3.2 Finite Element Theory

The majority of finite element codes for structural problems base their formulation on minimum potential energy. This means that the primary unknown field for the solid continuum is the displacement $\mathbf{u}$, whereas strains and stresses, $\boldsymbol{\epsilon}$ and $\boldsymbol{\sigma}$, respectively, are derived quantities. The existence of the variational principle further requires the forces to be derived from a potential, and when this is the case minimum potential energy can be expressed as

$$\Pi = \frac{1}{2}\int_V \boldsymbol{\epsilon}^T\mathbf{C}\boldsymbol{\epsilon}\,dV - \int_V \mathbf{u}^T\mathbf{f}\,dV \tag{3.21}$$

$\mathbf{C}$ is the constitutive matrix and $\mathbf{f}$ represents volumetric forces.

The first variation of equation (3.21) (when nodal displacements $\mathbf{v}$ are the only unknowns) is often referred to as the virtual work equation:

$$\delta\Pi = \int_V \delta\boldsymbol{\epsilon}^T\mathbf{C}\boldsymbol{\epsilon}\,dV - \int_V \delta\mathbf{u}^T\mathbf{f}\,dV \tag{3.22}$$
$$= \delta\mathbf{v}^T\mathbf{F}_i - \delta\mathbf{v}^T\mathbf{F}_e$$
$$= 0$$

In the equation above one has added the requirement that $\delta\Pi = 0$ since the desired solution minimizes the potential of equation (3.21). The vanishing of the virtual work for any virtual displacement field is only satisfied when the internal $\mathbf{F}_i$ and external forces $\mathbf{F}_e$ are at equilibrium.

The minimum potential energy equation is only valid for purely elastic problems when the applied forces are derivable from a work potential. However, it's variation, the virtual work equation of (3.22) also represents the equilibrium equation for a much more general class of problems; it is also correct for plasticity, damping and nonconservative loading. For this reason, the virtual work equation, i.e., equilibrium equation of (3.22), will serve as the true starting point for mechanism simulations, wheres the minimum potential energy of (3.21) will serve more of a "supporting" role where it is convenient with regard to the elastic behavior of mechanism links.

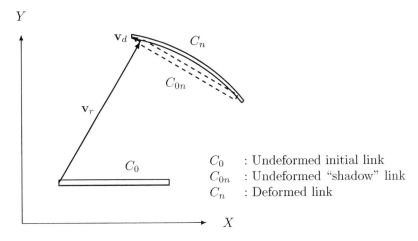

FIGURE 3.1: Co-rotational configurations

3.3 Corotational Formulation

We assume that each single link of the mechanism can be treated as an elastic body. Furthermore, we assume that within a *Local coordinate system*, the internal forces can be expressed through a linear relationship with respect to the *deformational displacements,* $\mathbf{v}_d$ of equation (3.23). This concept is the foundation for *corotational finite element formulations* [16, 44, 36].

For each link we picture an undeformed mesh that follows the link in such a way that the deformations (displacements from the undeformed mesh to the deformed mesh) are at a minimum. What this "minimum" position entails is in itself a topic of research, however, for small to moderate deformations the various methods do not alter the results perceivably, and more heuristic approaches to choosing the position of the undeformed body suffice.

The corotated undeformed mesh, or link, is in literature often referred to as the "Shadow element" or "Ghost reference" [5].

The total displacement, containing both rigid body motion $\mathbf{v}_r$ and deformations $\mathbf{v}_d$, is then split in two as

$$\mathbf{v} = \mathbf{v}_r + \mathbf{v}_d \tag{3.23}$$

For each point, or node when considering a discretized model, this separation can be defined when the rigid body motion has been defined. Figure 3.1 shows this decomposition of the displacements for a link.

3.4 Mechanism Links as Substructures and Super Elements

Performing a dynamic simulation for a mechanism often has a couple of seemingly competing characteristics:

1. For most mechanism structures, the number of connection points between the various links is quite small. Furthermore, the number of deformation modes being "activated" by a typical simulation is similarly small.

2. Stress concentrations around certain local geometric features of each link are often of great interest when trying to assess the viability of a mechanism; Fatigue life is highly dependent on accurate stress history predictions.

Item 1 suggests that mechanism analysis can be performed using relatively few degrees of freedom overall. However, 2 says that in order to obtain sufficient accuracy for the stress results, each link has to have a fine mesh, and thus many degrees of freedom. How can we achieve high stress results without increasing the number of degrees of freedom (too much)? The answer here is "Model Reduction", which in general can be expressed as

$$\mathbf{v} = \mathbf{Hq} \qquad (3.24)$$

$\mathbf{v}$ is the full set of degrees of freedom for the link mesh with size n, $\mathbf{q}$ is the reduced or condensed set of displacements of size m. One is looking to achieve $m \ll n$ without (too much of) a loss in accuracy for the stress results. Fortunately, for most mechanisms this is possible.

3.4.1 Static modes

When all loading and forced displacements are acting on the interface (external) degrees of freedom $\mathbf{v}_e$, being located at the triads, the internal degrees of freedom $\mathbf{v}_i$ can be expressed entirely as a function of the external degrees of freedom. This is provided that the velocities and accelerations are sufficiently low for the internal displacements of each link to be (close to) purely static. The stiffness relationship can then be expressed as

$$\begin{bmatrix} \mathbf{K}_{ii} & \mathbf{K}_{ie} \\ \mathbf{K}_{ei} & \mathbf{K}_{ee} \end{bmatrix} \begin{bmatrix} \mathbf{v}_i \\ \mathbf{v}_e \end{bmatrix} = \begin{bmatrix} \mathbf{Q}_i \\ \mathbf{Q}_e \end{bmatrix} \qquad (3.25)$$

where subscript i for internal and e for external degrees of freedom. Rewriting equation 1 of (3.25) gives us

$$\mathbf{v}_i = \mathbf{K}_{ii}^{-1}\mathbf{Q}_i - \mathbf{K}_{ii}^{-1}\mathbf{K}_{ie}\mathbf{v}_e \qquad (3.26)$$
$$= \mathbf{K}_{ii}^{-1}\mathbf{Q}_i + \mathbf{B}\mathbf{v}_e$$
$$= \mathbf{v}_i^i + \mathbf{v}_i^e$$

3.4.2   Fixed interface dynamic modes

When the mechanism is operating at high accelerations, inertia forces on the internal section start to influence the dynamics of the mechanism and hence also the stresses for the individual links. When this is the case, one can enhance the high frequency response and accuracy for the individual links by including internal dynamics of the link. The most common way of doing this is by including "fixed interface modes", often referred to as "Craig–Bampton modes" [4, 20]. When we fix the external degrees of freedom $\mathbf{v}_e$, the free undamped vibration of the internal degrees of freedom is given by

$$\mathbf{M}_{ii}\ddot{\mathbf{v}}_i + \mathbf{K}_{ii}\mathbf{v}_i = \mathbf{0} \tag{3.27}$$

When considering a simple harmonic motion expressed as

$$\mathbf{v}_i = \boldsymbol{\phi}\sin\omega t \tag{3.28}$$

we arrive at the eigenvalue problem

$$(\mathbf{K}_{ii} - \omega^2\mathbf{M}_{ii})\boldsymbol{\phi} = \mathbf{0} \tag{3.29}$$

Solving this eigenvalue problem for a selected number of eigenvalues and eigenvectors allows us to express the internal displacements as a linear combination of eigenvector amplitudes $\mathbf{y}$:

$$\mathbf{v}_i = \sum_{j=1}^{s} \boldsymbol{\phi}_j y_j = \boldsymbol{\Phi}\mathbf{y} \tag{3.30}$$

$$\text{where}\quad \boldsymbol{\phi} = \begin{bmatrix} \boldsymbol{\phi}_1 & \boldsymbol{\phi}_2 & \cdots & \boldsymbol{\phi}_s \end{bmatrix} \quad \text{and } \mathbf{y} = \begin{bmatrix} y_1 \\ y_2 \\ \vdots \\ y_s \end{bmatrix} \tag{3.31}$$

The number of selected eigenvectors s is usually much smaller than the number of internal degrees of freedom.

3.4.3   Reduced system

The superelement displacements are now expressed by the external DOFs $\mathbf{v}_e$ and by the new generalized DOFs $\mathbf{y}$:

$$\mathbf{v} = \begin{bmatrix} \mathbf{v}_e \\ \mathbf{v}_i \end{bmatrix} = \begin{bmatrix} \mathbf{I} & \mathbf{0} \\ \mathbf{B} & \boldsymbol{\Phi} \end{bmatrix} \begin{bmatrix} \mathbf{v}_e \\ \mathbf{y} \end{bmatrix} = \mathbf{H} \begin{bmatrix} \mathbf{v}_e \\ \mathbf{y} \end{bmatrix} \tag{3.32}$$

Usually only a few of the lowest modes of vibration need to be included to achieve good results, and this may reduce the size of the problem substantially. If all eigenmodes are included, $s = n-p$, the CMS (component mode synthesis) transformation is exact.

The substructure dynamic equation of motion may be written:

$$\begin{bmatrix} \mathbf{M}_{ee} & \mathbf{M}_{ei} \\ \mathbf{M}_{ie} & \mathbf{M}_{ii} \end{bmatrix} \begin{bmatrix} \ddot{\mathbf{v}}_e \\ \ddot{\mathbf{v}}_i \end{bmatrix} + \begin{bmatrix} \mathbf{C}_{ee} & \mathbf{C}_{ei} \\ \mathbf{C}_{ie} & \mathbf{C}_{ii} \end{bmatrix} \begin{bmatrix} \dot{\mathbf{v}}_e \\ \dot{\mathbf{v}}_i \end{bmatrix}$$
$$+ \begin{bmatrix} \mathbf{K}_{ee} & \mathbf{K}_{ei} \\ \mathbf{K}_{ie} & \mathbf{K}_{ii} \end{bmatrix} \begin{bmatrix} \mathbf{v}_e \\ \mathbf{v}_i \end{bmatrix} = \begin{bmatrix} \mathbf{Q}_e \\ \mathbf{Q}_i \end{bmatrix} \tag{3.33}$$

where $\mathbf{C}_{xx}$ represents damping.

Combining (3.32) and its first and second time derivatives with (3.33) and pre-multiplying with $\mathbf{H}^T$ produces

$$\begin{bmatrix} \mathbf{m}_{11} & \mathbf{m}_{12} \\ \mathbf{m}_{21} & \mathbf{m}_{22} \end{bmatrix} \begin{bmatrix} \ddot{\mathbf{v}}_e \\ \ddot{\mathbf{y}} \end{bmatrix} + \begin{bmatrix} \mathbf{c}_{11} & \mathbf{c}_{12} \\ \mathbf{c}_{21} & \mathbf{c}_{22} \end{bmatrix} \begin{bmatrix} \dot{\mathbf{v}}_e \\ \dot{\mathbf{y}} \end{bmatrix}$$
$$+ \begin{bmatrix} \mathbf{k}_{11} & \mathbf{k}_{12} \\ \mathbf{k}_{21} & \mathbf{k}_{22} \end{bmatrix} \begin{bmatrix} \mathbf{v}_e \\ \mathbf{y} \end{bmatrix} = \begin{bmatrix} \mathbf{q}_1 \\ \mathbf{q}_2 \end{bmatrix} \tag{3.34}$$

where

$$\begin{aligned} \mathbf{m}_{11} &= \mathbf{M}_{ee} + \mathbf{B}^T\mathbf{M}_{ie} + \mathbf{M}_{ei}\mathbf{B} + \mathbf{B}^T\mathbf{M}_{ii}\mathbf{B} & (3.35) \\ \mathbf{m}_{12} &= \mathbf{m}_{21}^T = \mathbf{M}_{ei}\boldsymbol{\Phi} + \mathbf{B}^T\mathbf{M}_{ii}\boldsymbol{\Phi} \\ \mathbf{m}_{22} &= \boldsymbol{\Phi}^T\mathbf{M}_{ii}\boldsymbol{\Phi} = \mathbf{I} \end{aligned}$$

$$\begin{aligned} \mathbf{c}_{11} &= \mathbf{C}_{ee} + \mathbf{B}^T\mathbf{C}_{ie} + \mathbf{C}_{ei}\mathbf{B} + \mathbf{B}^T\mathbf{C}_{ii}\mathbf{B} & (3.36) \\ \mathbf{c}_{12} &= \mathbf{c}_{21}^T = \mathbf{C}_{ei}\boldsymbol{\Phi} + \mathbf{B}^T\mathbf{C}_{ii}\boldsymbol{\Phi} \\ \mathbf{c}_{22} &= \boldsymbol{\Phi}^T\mathbf{C}_{ii}\boldsymbol{\Phi} \end{aligned}$$

$$\begin{aligned} \mathbf{k}_{11} &= \mathbf{K}_{ee} + \mathbf{K}_{ie}^T\mathbf{B} & (3.37) \\ \mathbf{k}_{12} &= \mathbf{k}_{21}^T = \mathbf{0} \\ \mathbf{k}_{22} &= \boldsymbol{\Phi}^T\mathbf{K}_{ii}\boldsymbol{\Phi} \end{aligned}$$

$$\begin{aligned} \mathbf{q}_1 &= \mathbf{Q}_e + \mathbf{B}^T\mathbf{Q}_i & (3.38) \\ \mathbf{q}_2 &= \boldsymbol{\Phi}^T\mathbf{Q}_i \end{aligned}$$

The matrix $\mathbf{m}_{22}$ from (3.35) is diagonal and with the eigenmodes being mass normalized, we have

$$\mathbf{m}_{22} = \boldsymbol{\Phi}^T\mathbf{M}_{ii}\boldsymbol{\Phi} = \mathbf{I} \tag{3.39}$$

Note that $\mathbf{m}_{11}$, $\mathbf{m}_{12}$, and $\mathbf{m}_{21}$ are not diagonal.

The matrix $\mathbf{k}_{11}$ from (3.37) is reduced from the expression

$$\mathbf{k}_{11} = \mathbf{K}_{ee} + \mathbf{B}^T\mathbf{K}_{ie} + \mathbf{K}_{ei}\mathbf{B} + \mathbf{B}^T\mathbf{K}_{ii}\mathbf{B} \tag{3.40}$$

Expanding the last two terms by (3.40) shows that the terms are equal, but have opposite signs; therefore, they cancel out of the equation. The similarity of these terms is due to the symmetric properties of the stiffness matrix, which produce

$$\mathbf{K}_{ie}^T = \mathbf{K}_{ei} \tag{3.41}$$

and

$$\left(\mathbf{K}_{ii}^{-1}\right)^T = \mathbf{K}_{ii}^{-1} \tag{3.42}$$

For the same reasons, the matrices $\mathbf{k}_{12} = \mathbf{k}_{21}^T$ are reduced from the equation

$$
\begin{aligned}
\mathbf{k}_{12} &= \mathbf{k}_{21}^T = \mathbf{K}_{ei}\boldsymbol{\Phi} + \mathbf{B}^T\mathbf{K}_{ii}\boldsymbol{\Phi} \\
&= \mathbf{K}_{ei}\boldsymbol{\Phi} + (-\mathbf{K}_{ii}^{-1}\mathbf{K}_{ie})^T\mathbf{K}_{ii}\boldsymbol{\Phi} \\
&= \mathbf{K}_{ei}\boldsymbol{\Phi} - \mathbf{K}_{ie}^T\left(\mathbf{K}_{ii}^{-1}\right)^T\mathbf{K}_{ii}\boldsymbol{\Phi} = \mathbf{0}
\end{aligned}
\tag{3.43}
$$

It may be shown that the stiffness matrix $\mathbf{k}_{22}$ is diagonal and of a form in which

$$\mathbf{k}_{22} = \begin{bmatrix} \omega_1^2 & \omega_2^2 & \cdots & \omega_{n-p}^2 \end{bmatrix} \tag{3.44}$$

ω_1^2, ω_2^2, ... ω_{n-p}^2 are the eigenvalues corresponding to the eigenmodes of eigenvector matrix $\boldsymbol{\Phi}$.

The substructure matrices reduced by CMS transformation will later form the basis for the mechanism simulation formulation.

3.4.4 Structural damping

The superelement damping matrix, given by equation (3.36) is not explicitly developed. Instead, for each element individually, proportional damping is used. If we assume that the damping force in a superelement is proportional to the velocity of each mass point, we have

$$\mathbf{c} = \alpha_1 \mathbf{m} \tag{3.45}$$

where α_1 is a constant. Similarly, if we assume the damping force is proportional to the strain velocity in each point, we have

$$\mathbf{c} = \alpha_2 \mathbf{k} \tag{3.46}$$

where α_2 is another constant. The combination of these two assumptions produces the damping matrix of *Rayleigh-damping* or *proportional damping*

$$\mathbf{c} = \alpha_1 \mathbf{m} + \alpha_2 \mathbf{k} \tag{3.47}$$

The damping ratio for the natural frequencies can now be calculated from

$$\lambda_i = \frac{1}{2}\left(\frac{\alpha_1}{\omega_i} + \alpha_2\omega_i\right) \tag{3.48}$$

where α_1 damps out lower vibration modes while α_2 damps out higher modes.

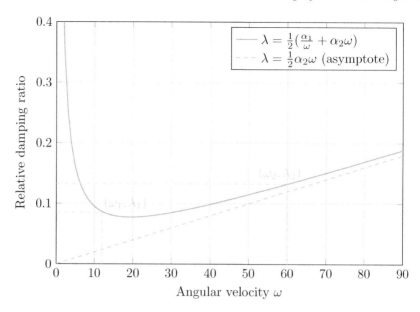

FIGURE 3.2: Rayleigh damping: Typical relationship between damping and natural frequency arising from the specification of damping ratio at two particular frequencies

If the damping ratios λ_i for two vibration modes are selected, the corresponding constants of proportionality, α_1 and α_2 may be calculated from

$$\alpha_1 = \frac{2\omega_1\omega_2}{\omega_2^2 - \omega_1^2}\left(\lambda_1\omega_2 - \lambda_2\omega_1\right) \tag{3.49}$$

$$\alpha_2 = \frac{2\left(\omega_2\lambda_2 - \omega_1\lambda_1\right)}{\omega_2^2 - \omega_1^2} \tag{3.50}$$

where ω_1 and ω_2 are the circle frequencies and λ_1 and λ_2 are the damping ratios for the selected vibration modes; see Figure 3.2.

When using component modes (see Section 3.4), the reduced superelement mass- and stiffness matrices are partitioned as given by equations (3.35) and (3.37). It is then possible to assign individual Rayleigh damping factors for each component mode. In this case, exploiting that $\mathbf{k}_{12} = \mathbf{k}_{21}^T = \mathbf{0}$ and $\mathbf{m}_{12} = \mathbf{m}_{21}^T$, equation (3.47) reads

$$\begin{bmatrix} \mathbf{c}_{11} & \mathbf{c}_{12} \\ \mathbf{c}_{21} & \mathbf{c}_{22} \end{bmatrix} = \begin{bmatrix} \alpha_1\mathbf{m}_{11} & \left(\boldsymbol{\alpha}_m\mathbf{m}_{21}\right)^T \\ \boldsymbol{\alpha}_m\mathbf{m}_{21} & \boldsymbol{\alpha}_m\mathbf{m}_{22} \end{bmatrix} + \begin{bmatrix} \alpha_2\mathbf{k}_{11} & \mathbf{0} \\ \mathbf{0} & \boldsymbol{\alpha}_k\mathbf{k}_{22} \end{bmatrix} \tag{3.51}$$

where $\boldsymbol{\alpha}_m = \lceil\alpha_{mi}\rfloor$ and $\boldsymbol{\alpha}_k = \lceil\alpha_{ki}\rfloor$ are diagonal matrices containing the component mode damping factors. Note that $\mathbf{m}_{22}$ and $\mathbf{k}_{22}$ both are diagonal matrices; see Section 3.4.3.

3.5 FE Modeling of Joints

3.5.1 Description of joint motion

The motion of the mechanism links is governed by various types of joints between them. The relative motion between two links can be described through a series of relative motions using homogeneous transformations. We designate one link as the master link, and the positions of its joint attachment (master triad) point by the 4×4 matrix $\mathbf{P}_{MG}$. On the other side of the joint, we have the slave link with its attachment point (slave triad) $\mathbf{P}_{SG}$. Subscript G is used to reference global coordinate system. The overall relative position of the slave triad in global system, $\mathbf{P}_{SG}$ can be described as

$$\mathbf{P}_{SG} = \mathbf{P}_{MG}\mathbf{P}_{JM}\mathbf{P}_{J}\mathbf{P}_{SJ} \tag{3.52}$$

The actual motion due to joint degrees of freedom (DOFs) is described by the relative motion $\mathbf{P}_J$. This can consist of a number of relative motions

$$\mathbf{P}_J = \mathbf{P}_{J1}\mathbf{P}_{J2}\cdots\mathbf{P}_{JN} \tag{3.53}$$

Equation (3.52) is derived by starting from the master triad position $\mathbf{P}_{MG}$ and consecutively adding the relative motions, or positions, of the joint until we arrive at the position of the slave triad $\mathbf{P}_{SG}$, measured in the global coordinate system.

The different terms of equation (3.52) are:

$\mathbf{P}_{SG}$ Position of slave triad measured in global coordinate system. The position is the product of all the joint degrees of freedom (DOFs).

$\mathbf{P}_{MG}$ Position of the master triad measured in global coordinate system.

$\mathbf{P}_{JM}$ Position of the joint relative to the master triad, measure in the master triad coordinate system.

$\mathbf{P}_J$ Relative motion through the joint itself. When all joint DOFs are zero, this will reduce to identity, i.e., no relative motion.

$\mathbf{P}_{SJ}$ Position of slave triad relative to the joint, measured in joint coordinate system.

3.5.1.1 Variation of joint motions

In order to perform the dynamic simulation, we need the variation of the slave triad position and orientation with respect to all its master degrees of freedom; master triad degrees of freedom and joint degrees of freedom of (3.52). This variation can be expressed as

$$\delta \mathbf{P}_{SG} = \delta \mathbf{P}_{MG} \mathbf{P}_{JT} \mathbf{P}_N \cdots \mathbf{P}_1 \mathbf{P}_{SJ} \qquad (3.54)$$
$$+ \mathbf{P}_{MG} \mathbf{P}_{JT} \delta \mathbf{P}_N \cdots \mathbf{P}_1 \mathbf{P}_{SJ}$$
$$\vdots$$
$$+ \mathbf{P}_{MG} \mathbf{P}_{JT} \mathbf{P}_N \cdots \delta \mathbf{P}_1 \mathbf{P}_{SJ}$$

In the equation above one has utilized the chain rule during variation, keeping in mind that $\mathbf{P}_{JT}$ and $\mathbf{P}_{SJ}$ are constant and thus $\delta \mathbf{P}_{JT} = \mathbf{0}$ and $\delta \mathbf{P}_{SJ} = \mathbf{0}$.

Variation of the master triad degrees of freedom gives us

$$\begin{bmatrix} \delta \mathbf{u}_s \\ \delta \omega_s \end{bmatrix} = \begin{bmatrix} \mathbf{I} & \widehat{\mathbf{e}}_{SM} \\ \mathbf{0} & \mathbf{I} \end{bmatrix} \begin{bmatrix} \delta \mathbf{u}_m \\ \delta \omega_m \end{bmatrix} \qquad (3.55)$$

$\mathbf{e}_{SM}$ is the position of the slave triad relative to the master triad, and $\widehat{\mathbf{e}}_{SM}$ is the skew-symmetric spin matrix defined according to (3.18).

The variation of the joint degrees of freedom of a particular matrix $\mathbf{P}_i$ is similarly given by

$$\begin{bmatrix} \delta \mathbf{u}_s \\ \delta \omega_s \end{bmatrix} = \begin{bmatrix} \mathbf{I} & \widehat{\mathbf{e}}_{SM} \\ \mathbf{0} & \mathbf{I} \end{bmatrix} \begin{bmatrix} \mathbf{R}_{C_i} & \mathbf{0} \\ \mathbf{0} & \mathbf{R}_{C_i} \end{bmatrix} \begin{bmatrix} \mathbf{0} & \mathbf{0} \\ \mathbf{0} & \mathbf{H}_i \end{bmatrix} \begin{bmatrix} \delta \mathbf{u}_i \\ \delta \omega_i \end{bmatrix} \qquad (3.56)$$
$$= \begin{bmatrix} \mathbf{R}_{C_i} & \widehat{\mathbf{e}}_{SM} \mathbf{R}_{C_i} \mathbf{H}_i \\ \mathbf{0} & \mathbf{R}_{C_i} \mathbf{H}_i \end{bmatrix} \begin{bmatrix} \delta \mathbf{u}_i \\ \delta \omega_i \end{bmatrix}$$

3.6 Joint Specializations

As can be seen, the general form of (3.52) only needs the specializations of (3.53) to describe a wide range of joint types

3.6.1 Rigid joint

A rigid joint fixes the position of the slave node relative to the master node and one thus have

$$\mathbf{P}_J = \mathbf{I} \qquad (3.57)$$

The joint represents a convenient way of locking the relative position of two nodes that do not have to be at the same location.

3.6.2 Revolute, cylindric and prismatic joint with a single master

The revolute joint is described by using one relative motion matrix with degree of freedom being the rotation $\tilde{\theta}_z$ about the local Z axis of the joint; see

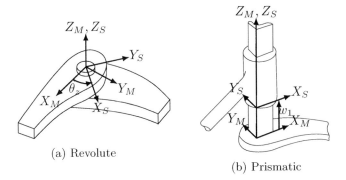

(a) Revolute

(b) Prismatic

FIGURE 3.3: Revolute and prismatic joint degrees of freedom

Figure 3.3.

$$\mathbf{P}_J = \mathbf{P}_{J1}(\tilde{\theta}_z) \tag{3.58}$$

When adding a translational degree of freedom along the Z axis in the same position matrix, one arrives at a cylindrical joint with a single master node.

$$\mathbf{P}_J = \mathbf{P}_{J1}(\tilde{\theta}_z, \tilde{w}_z) \tag{3.59}$$

With only the translational degree of freedom along the Z axis, one arrives at the formulation for the prismatic joint:

$$\mathbf{P}_J = \mathbf{P}_{J1}(\tilde{w}_z) \tag{3.60}$$

3.6.3 Universal joint

A universal joint (see Figure 3.4) is defined by two consecutive rotations

$$\mathbf{P}_J = \mathbf{P}_{J1}(\tilde{\theta}_z)\mathbf{P}_{J2}(\tilde{\theta}_y) \tag{3.61}$$

The first rotation $\mathbf{P}_{J1}(\tilde{\theta}_z)$ rotates the cross about the fixed axis Z of the master side of the joint. The second rotation $\mathbf{P}_{J2}(\tilde{\theta}_y)$ rotates the slave side of the joint an angle $\tilde{\theta}_y$ about the Y axis of the cross.

When the input axis and output axis of a universal joint has a non-zero angle, the rotational velocity of the output axis oscillates relative to the input rotational velocity.

3.6.4 Constant velocity joint

A common way of constructing a constant velocity joint is to combine two universal joints where the axis angle of the first and second joint is constrained to be equal to each other; i.e., the total axis angle is halved across the two

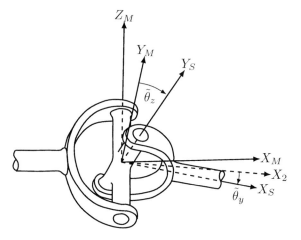

FIGURE 3.4: Universal joint coordinate systems and degrees of freedom

universal joints. This approach is common in for instance wide angle power take off (PTO) shafts for tractors.

This approach can be used for formulating a constant velocity joint as well. Using 4 relative motion matrices and constraining the rotation of the first and last matrix to have the same rotation, and also constraining the relative rotation of the second and third matrix to be the same, one arrives at the joint formulation

$$\mathbf{P}_J = \mathbf{P}_{J1}(\tilde{\theta}_z)\mathbf{P}_{J2}(\tilde{\theta}_y)\mathbf{P}_{J3}(\tilde{\theta}_y)\mathbf{P}_{J2}(\tilde{\theta}_z) \qquad (3.62)$$

Note that there are only two joint degrees of freedom; $\tilde{\theta}_z$ and $\tilde{\theta}_y$ but they each appear in two matrices.

3.6.5   Multimaster joints

Cylindrical and prismatic joints with a limited motion can give sufficient accuracy with a single master node as described in (3.59) when the axial displacement is small relative to the master. However, for a number of cases this does not give sufficient accuracy: when the motion is relatively large, the sliding axis part is relatively flexible; one wants to accurately predict the stresses along the sliding axis part. When any of these is the case, improved accuracy is achieved by describing the motion in the axial direction as the interpolation of a number of master nodes along the slider; see Figure 3.5. By defining the slider variable s as the position along the slider, the joint position can be described in general as an interpolation of the discrete joint position at each

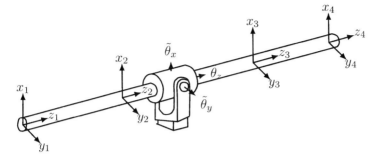

FIGURE 3.5: Cylindric multimaster joint

node i; $\mathbf{P}_{JG_i}$ of the total N number of master nodes.

$$\mathbf{P}_{JG} = \mathbf{P}(s, \mathbf{P}_{JG_1}, \cdots, \mathbf{P}_{JG_N})$$

$$= \sum_{i=1}^{N} f(s)_i \mathbf{P}_{JG_i} \quad \text{where} \quad \mathbf{P}_{JG_i} = \mathbf{P}_{MG_i} \mathbf{P}_{JM_i} \tag{3.63}$$

With a linear interpolation between the master triads, and the follower positioned between node k and l, the interpolation functions $f(s)_i$ will be

$$\begin{aligned} f(s)_i &= (-s - s_l)/(s_l - s_k) \\ f(s)_l &= (s - s_k)/(s_l - s_k) \\ f(s)_i &= 0 \quad \text{for } i \neq k \text{ and } i \neq l \end{aligned} \tag{3.64}$$

which resembles the similar expression for the single master joint.

The variation of (3.63) with respect to the N master triads is

$$\begin{bmatrix} \delta\mathbf{u}_s \\ \delta\boldsymbol{\omega}_s \end{bmatrix}_M = \sum_{i=1}^{N} f(s)_i \begin{bmatrix} \mathbf{I} & \hat{\mathbf{e}}_{SM_i} \\ \mathbf{0} & \mathbf{I} \end{bmatrix} \begin{bmatrix} \delta\mathbf{u}_{m_i} \\ \delta\boldsymbol{\omega}_{m_i} \end{bmatrix} \tag{3.65}$$

Prismatic joint has a fixed orientation about the slider axis (z-axis), and in this regard resembles the single master prismatic joint of (3.60). In addition the joint has a free rotations about the x- and y-axis of the slider. This gives the joint motion parameterization

$$\mathbf{P}_J = \mathbf{P}_{J2}(\tilde{\theta}_y)\mathbf{P}_{J1}(\tilde{\theta}_x) \tag{3.66}$$

Cylindric joint further has a free rotation about the slider z-axis

$$\mathbf{P}_J = \mathbf{P}_{J3}(\tilde{\theta}_z)\mathbf{P}_{J2}(\tilde{\theta}_y)\mathbf{P}_{J1}(\tilde{\theta}_x) \tag{3.67}$$

3.6.6 Transmissions between joint variables

The most common transmissions can be defined through linear constraints between joint variables of different joints.

Gear transmission is defined by making the output rotation $\tilde{\theta}_{z_O}$ dependent on the input rotation $\tilde{\theta}_{z_I}$ through the gear ratio R

$$\tilde{\theta}_{z_O} = R\,\tilde{\theta}_{z_I} \tag{3.68}$$

Rack and pinion is defined by tying the output axial displacement of a prismatic joint to the rotational degree of freedom of an input revolute joint

$$s_O = R\,\tilde{\theta}_{z_I} \tag{3.69}$$

Screw joint follows the same linear relationship as the rack and pinion, except that the input rotation is the $\tilde{\theta}_z$ of a cylindric joint and the output is the slider variable of the same cylindric joint.

$$s_O = R\,\tilde{\theta}_{z_I} \tag{3.70}$$

3.7 Multipoint Constraints

The point to point joints are connected to single nodes at two different element meshes (parts). When refining a mesh around a single joint connecting node, the strains and stresses around this node will approach infinity if the elements attached to the node are only shell or solid elements. This behavior described by Fridel Hartmann [19].

In order to make the joint attachment more physically correct, one should connect the attachment point to a number of neighboring points. One remedy to this is to attach beam elements between the joint attachment point and a number of neighboring nodes, but a more elegant approach is often to apply either a rigid attachment to the neighboring nodes, or define the joint connection node as a weighted average of a number of neighboring nodes.

3.7.1 Rigid multipoint constraint

Rigid multipoint constraints are defined by having a number of slave nodes being rigidly attached to a single master node; see Figure 3.6. If we denote each slave node position relative to the master node by $\mathbf{e}_i$, the incremental (linearized) translation of the slave node $\delta\mathbf{u}_{si}$ can be expressed with respect to the master translation $\delta\mathbf{u}_s$ and rotation $\delta\boldsymbol{\omega}_m$

$$\begin{bmatrix} \delta\mathbf{u}_{si} \\ \delta\boldsymbol{\omega}_{s}i \end{bmatrix} = \begin{bmatrix} \mathbf{I} & \widehat{\mathbf{e}}_i \\ \mathbf{0} & \mathbf{I} \end{bmatrix} \begin{bmatrix} \delta\mathbf{u}_m \\ \delta\boldsymbol{\omega}_m \end{bmatrix} \quad \text{where} \quad \widehat{\mathbf{e}}_i = \begin{bmatrix} 0 & -e_z & e_y \\ e_z & 0 & -e_x \\ -e_y & e_x & 0 \end{bmatrix} \tag{3.71}$$

This type of rigid connection has different names in various CAE tools, and is for instance referred to as RBE2 in MSC Nastran [2].

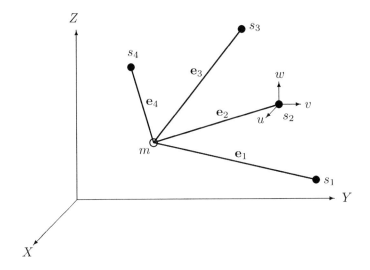

FIGURE 3.6: Multinode rigid element

3.7.2 Weighted average multipoint constraint

The flexible multi point constraint will describe the motion of the slave node as a weighted average of a number of master nodes; see Figure 3.7.

The weighting factor for each master node j is denoted w_{j_i} where index i represents displacement component in x, y and z for the values 1, 2, and 3. The weighting factors for rotations ($i \in 4, 5, 6$) follow the weighting factors for translations with the additional factor of squared length between the node and centroid of the masters.

$$w_{j_{i+3}} = w_{j_i} l_j^2 \qquad (3.72)$$

The weight factors for the nodal degrees of freedom for node j can then be collected in a diagonal matrix:

$$\mathbf{W}_j = \lceil \; w_{j1} \quad w_{j2} \quad w_{j3} \quad w_{j4} \quad w_{j5} \quad w_{j6} \; \rfloor \qquad (3.73)$$

If a load at the slave node is applied, the effect is that this load is distributed to all the master nodes with respect to their mutual weighting factors. The flexible multipoint constraint does not constrain the mutual motion of the master nodes; the only constraining effect comes from the force equilibrium between the master nodes and the slave nodes.

If a force $\mathbf{f}_j$ and moment $\mathbf{m}_j$ is acting on a master node j, its statically equivalent load $\mathbf{f}_s$ and moment $\mathbf{m}_s$ at the dependent node s can be expressed as

$$\begin{bmatrix} \mathbf{f}_s \\ \mathbf{m}_s \end{bmatrix} = \begin{bmatrix} \mathbf{I} & \mathbf{0} \\ \widehat{\mathbf{e}}_j & \mathbf{I} \end{bmatrix} \begin{bmatrix} \mathbf{f}_j \\ \mathbf{m}_j \end{bmatrix} \qquad \text{or} \qquad \mathbf{F}_s = \mathbf{E}_j^T \mathbf{F}_j \qquad (3.74)$$

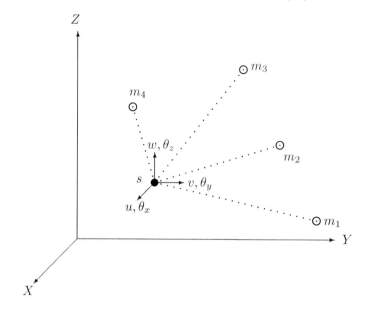

FIGURE 3.7: Multinode weighted averaged motion element

The sum of all contributions from the master nodes is then

$$\mathbf{F}_s = \sum_j \mathbf{E}_j^T \mathbf{F}_j \tag{3.75}$$

If the dependent node s is subject to a force and moment $\mathbf{F}_s$, we assume that we can distribute this force to the master nodes using the weighting factors $\mathbf{W}_j$ from (3.73) and the eccentricities $\mathbf{E}_j$ from (3.74) as:

$$\mathbf{F}_j = \mathbf{W}_j \mathbf{E}_j \mathbf{X} \mathbf{F}_s \tag{3.76}$$

The 6×6 matrix $\mathbf{X}$ is the same for all nodes j. By combining (3.75) and (3.76) we arrive at the identity:

$$\mathbf{F}_s = \sum_j \mathbf{E}_j^T \mathbf{F}_j = \sum_j \mathbf{E}_j^T \mathbf{W}_j \mathbf{E}_j \mathbf{X} \mathbf{F}_s = \mathbf{I} \, \mathbf{F}_s \tag{3.77}$$

For this to be true, we must have

$$\mathbf{X} = \left(\sum_j \mathbf{E}_j^T \mathbf{W}_j \mathbf{E}_j \right)^{-1} \tag{3.78}$$

Having arrived at the expression for $\mathbf{X}$, the initial assumed form (3.76) can now be used to define the contribution to a master node j from a load at the dependent node s

$$\mathbf{F}_j = \mathbf{G}_j \mathbf{F}_s \qquad \text{where} \qquad \mathbf{G}_j = \mathbf{W}_j \mathbf{E}_j \mathbf{X} \tag{3.79}$$

$\mathbf{G}_j$ is then a 6×6 distribution matrix giving the contribution to master node j.

So far we have looked at the force distribution problem, and the principle of virtual work gives us the following equation

$$\delta \mathbf{v}_s{}^T \mathbf{F}_s = \sum_j \delta \mathbf{v}_j^T \mathbf{F}_s = \sum_j \delta \mathbf{v}_j^T \mathbf{G}_j \mathbf{F}_s \tag{3.80}$$

For this to be satisfied for all $\mathbf{F}_s$, the following relationship emerges regarding the displacement of the dependent node s

$$\delta \mathbf{v}_s = \sum_j \mathbf{G}_j^T \delta \mathbf{v}_j \tag{3.81}$$

3.8 Time Integration Methods for Nonlinear Dynamics

A mechanism going through motions and displacements will at any given time satisfy the dynamic equilibrium equation; Newton's second law. For the sake of being solved through the use of numerical time integration, Newton's second law is presented as a force residual for each degree of freedom of the system being equal to zero at all times.

$$\mathbf{R}(\mathbf{v}, \dot{\mathbf{v}}, \ddot{\mathbf{v}}, t) = \mathbf{0} \tag{3.82}$$

The residual equation can be viewed as the sum of inertia, damping, deformational and applied forces.

$$\mathbf{R}(\mathbf{v}, \dot{\mathbf{v}}, \ddot{\mathbf{v}}, t) = \mathbf{p}(\mathbf{v}, \dot{\mathbf{v}}, \ddot{\mathbf{v}}, t) - \mathbf{f}(\mathbf{v}, t) = \mathbf{F}^I + \mathbf{F}^D + \mathbf{F}^S - \mathbf{Q} = \mathbf{0} \tag{3.83}$$

The terms inertia, damping and deformational forces become more apparent when one introduces the first variation of the residual equation (3.82). The dynamic solution of the elastic mechanism is a nonlinear path through the solutions space being parameterized with respect to time. Since (3.82) has to be satisfied at all times, the following variation is also satisfied.

$$\begin{aligned} \delta \mathbf{R} &= \frac{\partial \mathbf{R}}{\partial \ddot{\mathbf{v}}} \delta \ddot{\mathbf{v}} + \frac{\partial \mathbf{R}}{\partial \dot{\mathbf{v}}} \delta \dot{\mathbf{v}} + \frac{\partial \mathbf{R}}{\partial \mathbf{v}} \delta \mathbf{v} + \frac{\partial \mathbf{R}}{\partial t} \delta t \\ &= \mathbf{M} \delta \ddot{\mathbf{v}} + \mathbf{C} \delta \dot{\mathbf{v}} + \mathbf{K} \delta \mathbf{v} - \mathbf{q} \delta t \\ &= \delta \mathbf{F}^I + \delta \mathbf{F}^D + \delta \mathbf{F}^S - \delta \mathbf{Q} \\ &= \mathbf{0} \end{aligned} \tag{3.84}$$

Note that the variations $\delta\ddot{\mathbf{v}}$, $\delta\dot{\mathbf{v}}$, $\delta\mathbf{v}$, and δt are not independent, but mutually constrained to satisfy a kinematically admissible variation of displacements and velocities.

Using the notation $\Delta\ddot{\mathbf{v}} = \frac{\partial\ddot{\mathbf{v}}}{\partial t}\Delta t$, $\Delta\dot{\mathbf{v}} = \frac{\partial\dot{\mathbf{v}}}{\partial t}\Delta t$, and $\Delta\mathbf{v} = \frac{\partial\mathbf{v}}{\partial t}\Delta t$ we arrive at the incremental residual equation.

For a finite time-step Δt moving from one step to the next, we have

$$\begin{aligned} \mathbf{M}\Delta\ddot{\mathbf{v}} + \mathbf{C}\Delta\dot{\mathbf{v}} + \mathbf{K}\Delta\mathbf{v} - \mathbf{q}\Delta t &= 0 \\ \Delta\mathbf{F}^I + \Delta\mathbf{F}^D + \Delta\mathbf{F}^S - \Delta\mathbf{Q} &= 0 \end{aligned} \qquad (3.85)$$

$$\text{where} \quad \begin{cases} \Delta\mathbf{F}^I = \mathbf{M}\Delta\ddot{\mathbf{v}} & : \text{inertia forces} \\ \Delta\mathbf{F}^D = \mathbf{C}\Delta\dot{\mathbf{v}} & : \text{damping forces} \\ \Delta\mathbf{F}^S = \mathbf{K}\Delta\mathbf{v} & : \text{internal forces} \\ \Delta\mathbf{Q} = \mathbf{q}\Delta t & : \text{external forces} \end{cases} \qquad (3.86)$$

3.8.1 Newmark integration

The basis for the Newmark integration is the update equations for displacement and velocity

$$\mathbf{v}_{k+1} = \mathbf{v}_k + h\dot{\mathbf{v}}_k + h^2(\frac{1}{2} - \beta)\ddot{\mathbf{v}}_k + h^2\beta\ddot{\mathbf{v}}_{k+1} \qquad (3.87)$$

$$\dot{\mathbf{v}}_{k+1} = \dot{\mathbf{v}}_k + h(1 - \gamma)\ddot{\mathbf{v}}_k + h\gamma\ddot{\mathbf{v}}_{k+1} \qquad (3.88)$$

where subscript k signifies timestep.

In terms of propagating the solution for a range of algorithms, (3.88) can be written in an incremental form

$$\mathbf{v}_{k+1} = \mathbf{v}_k + \Delta\mathbf{v}_k \qquad \text{where} \quad \Delta\mathbf{v}_k = h\dot{\mathbf{v}}_k + h^2\frac{1}{2}\ddot{\mathbf{v}}_k + h^2\beta\Delta\ddot{\mathbf{v}}_k \qquad (3.89)$$

$$\dot{\mathbf{v}}_{k+1} = \dot{\mathbf{v}}_k + \Delta\dot{\mathbf{v}}_k \qquad \text{where} \quad \Delta\dot{\mathbf{v}}_k = h\ddot{\mathbf{v}}_k + h\gamma\Delta\ddot{\mathbf{v}}_k \qquad (3.90)$$

and $\Delta\ddot{\mathbf{v}}_k = \ddot{\mathbf{v}}_{k+1} - \ddot{\mathbf{v}}_k$.

3.8.1.1 Newmark with respect to displacement increment

When incrementing the solution, we want to be able to solve the equation system with respect to displacement increment $\Delta\mathbf{v}_k$. From (3.89) one gets the acceleration increment $\Delta\ddot{\mathbf{v}}$, and further combining (3.91) with (3.90) gives the velocity increment $\Delta\dot{\mathbf{v}}$, both expressed in terms of the displacement increment $\Delta\mathbf{v}$

$$\Delta\ddot{\mathbf{v}}_k = \frac{1}{h^2\beta}\Delta\mathbf{v}_k - \frac{1}{h\beta}\dot{\mathbf{v}}_k - \frac{1}{2\beta}\ddot{\mathbf{v}}_k \quad = \frac{1}{h^2\beta}\Delta\mathbf{v}_k - \mathbf{a}_k \qquad (3.91)$$

$$\Delta\dot{\mathbf{v}}_k = \frac{\gamma}{h\beta}\Delta\mathbf{v}_k - \frac{\gamma}{\beta}\dot{\mathbf{v}}_k - h(\frac{\gamma}{2\beta} - 1)\ddot{\mathbf{v}}_k = \frac{\gamma}{h\beta}\Delta\mathbf{v}_k - \mathbf{d}_k \qquad (3.92)$$

By inserting equations for incremental velocity (3.92) and incremental acceleration (3.91) into the residual equation on incremental form (3.86) one gets

$$\mathbf{M}(\frac{1}{h^2\beta}\Delta\mathbf{v}_k - \frac{1}{h\beta}\dot{\mathbf{v}}_k - \frac{1}{2\beta}\ddot{\mathbf{v}}_k) \tag{3.93}$$

$$+\mathbf{C}(\frac{\gamma}{h\beta}\Delta\mathbf{v}_k - \frac{\gamma}{\beta}\dot{\mathbf{v}}_k - h(\frac{\gamma}{2\beta}-1)\ddot{\mathbf{v}}_k) + \mathbf{K}\Delta\mathbf{v}_k - \Delta\mathbf{Q}_k = 0$$

Collecting all terms with unknown displacement increment on the left-hand side, and all known quantities on the right-hand side gives us

$$\mathbf{N}\Delta\mathbf{v}_k = \Delta\widehat{\mathbf{Q}} \tag{3.94}$$

where

$$\mathbf{N} = \frac{1}{h^2\beta}\mathbf{M} + \frac{\gamma}{h\beta}\mathbf{C} + \mathbf{K} \tag{3.95}$$

$$\Delta\widehat{\mathbf{Q}} = \Delta\mathbf{Q}_k + \mathbf{M}\mathbf{a}_k + \mathbf{C}\mathbf{d}_k \tag{3.96}$$

$$\mathbf{a}_k = \frac{1}{h\beta}\dot{\mathbf{v}}_k + \frac{1}{2\beta}\ddot{\mathbf{v}}_k \tag{3.97}$$

$$\mathbf{d}_k = \frac{\gamma}{\beta}\dot{\mathbf{v}}_k + h(\frac{\gamma}{2\beta}-1)\ddot{\mathbf{v}}_k \tag{3.98}$$

3.8.2 Generalized α method

The generalized α method by Chung and Hulbert [9] seeks to introduce high frequency dissipation into the numerical solution by interpolating the inertia forces between timestep k and $k+1$ using a factor α_m and interpolating between elastic, damping, external forces using α_f. The residual equation of (3.82) then takes the form

$$\begin{aligned}\mathbf{R}_\alpha = \quad &(1-\alpha_m)\mathbf{F}^I_{k+1} + \alpha_m\mathbf{F}^I_k \\ +&(1-\alpha_f)\mathbf{F}^D_{k+1} + \alpha_f\mathbf{F}^D_k \\ +&(1-\alpha_f)\mathbf{F}^S_{k+1} + \alpha_f\mathbf{F}^S_k \\ -&((1-\alpha_f)\mathbf{q}_{k+1} + \alpha_f\mathbf{q}_k) = 0\end{aligned} \tag{3.99}$$

Utilizing the variational expressions from (3.87) one obtains the incremental expressions

$$\mathbf{F}^I_{k+1} = \mathbf{F}^I_k + \frac{\partial\mathbf{R}}{\partial\ddot{\mathbf{v}}}\Delta\ddot{\mathbf{v}}_k = \mathbf{F}^I_k + \mathbf{M}\Delta\ddot{\mathbf{v}}_k \tag{3.100}$$

$$\mathbf{F}^D_{k+1} = \mathbf{F}^D_k + \frac{\partial\mathbf{R}}{\partial\dot{\mathbf{v}}}\Delta\dot{\mathbf{v}}_k = \mathbf{F}^D_k + \mathbf{C}\Delta\dot{\mathbf{v}}_k \tag{3.101}$$

$$\mathbf{F}^S_{k+1} = \mathbf{F}^S_k + \frac{\partial\mathbf{R}}{\partial\mathbf{v}}\Delta\mathbf{v}_k = \mathbf{F}^S_k + \mathbf{K}\Delta\mathbf{v}_k \tag{3.102}$$

using these expressions in (3.100) in incremental form gives us the generalized α-method equation on incremental form

$$(1 - \alpha_m)\mathbf{M}\Delta\ddot{\mathbf{v}}_k + \mathbf{F}_k^I \tag{3.103}$$

$$+(1 - \alpha_f)\mathbf{C}\Delta\ddot{\mathbf{v}}_k + \mathbf{F}_k^D \tag{3.104}$$

$$+(1 - \alpha_f)\mathbf{K}\Delta\dot{\mathbf{v}}_k + \mathbf{F}_k^S \tag{3.105}$$

$$-((1 - \alpha_f)\Delta\mathbf{q}_k + \mathbf{q}_k) = \mathbf{0} \tag{3.106}$$

Collecting all known terms, i.e., timestep k terms on the right-hand side gives us

$$(1 - \alpha_m)\mathbf{M}\Delta\ddot{\mathbf{v}}_k + (1 - \alpha_f)\mathbf{C}\Delta\ddot{\mathbf{v}}_{k+1} + (1 - \alpha_f)\mathbf{K}\Delta\dot{\mathbf{v}}_{k+1} \tag{3.107}$$

$$=(1 - \alpha_f)\Delta\mathbf{Q}_k - (\mathbf{F}_k^I + \mathbf{F}_k^D + \mathbf{F}_k^S - \mathbf{q}_k) \tag{3.108}$$

The last term on the right-hand side of (3.108) can be recognized as the residual equation at time k, and as such can be omitted since this should be $\mathbf{0}$ before stepping to $k + 1$. This interpolated equilibrium equation leads to the update equations.

3.8.3   Generalized α method with Newmark integration

Inserting the Newmark incremental acceleration and velocity expressed by displacement into the generalize α equation give us

$$(1 - \alpha_m)\mathbf{M}(\frac{1}{h^2\beta}\Delta\mathbf{v}_k - \frac{1}{h\beta}\dot{\mathbf{v}}_k - \frac{1}{2\beta}\ddot{\mathbf{v}}_k) \tag{3.109}$$

$$+(1 - \alpha_f)\mathbf{C}(\frac{\gamma}{h\beta}\Delta\mathbf{v}_k - \frac{\gamma}{\beta}\dot{\mathbf{v}}_k - h(\frac{\gamma}{2\beta} - 1)\ddot{\mathbf{v}}_k) \tag{3.110}$$

$$+(1 - \alpha_f)\mathbf{K}\Delta\mathbf{v}_k \tag{3.111}$$

$$=(1 - \alpha_f)\Delta\mathbf{Q}_k - (\mathbf{F}_k^I + \mathbf{F}_k^D + \mathbf{F}_k^S - \mathbf{Q}_k) \tag{3.112}$$

When collecting the unknown incremental displacements on the left-hand side, and known quantities on the right-hand side

$$\mathbf{N}\Delta\mathbf{v}_k = \Delta\widehat{\mathbf{Q}}_k \tag{3.113}$$

where

$$\mathbf{N} = \frac{(1 - \alpha_m)}{h^2\beta}\mathbf{M} + \frac{(1 - \alpha_f)\gamma}{h\beta}\mathbf{C} + (1 - \alpha_f)\mathbf{K} \tag{3.114}$$

$$\Delta\widehat{\mathbf{Q}}_k = (1 - \alpha_f)\mathbf{Q}_{k+1} + \alpha_f\mathbf{Q}_k \tag{3.115}$$

$$- (\mathbf{F}_k^I + \mathbf{F}_k^D + \mathbf{F}_k^S) + (\mathbf{M}\mathbf{a}_k + \mathbf{C}\mathbf{d}_k) \tag{3.116}$$

$$\mathbf{a}_k = \frac{1}{h\beta}\dot{\mathbf{v}}_k + \frac{1}{2\beta}\ddot{\mathbf{v}}_k \tag{3.117}$$

$$\mathbf{d}_k = \frac{\gamma}{\beta}\dot{\mathbf{v}}_k + h(\frac{\gamma}{2\beta} - 1)\ddot{\mathbf{v}}_k \tag{3.118}$$

Updating the state is done through equations (3.89) and (3.90), repeated for brevity:

$$\mathbf{v}_{k+1} = \mathbf{v}_k + \Delta\mathbf{v}_k \tag{3.119}$$

$$\dot{\mathbf{v}}_{k+1} = \dot{\mathbf{v}}_k + \Delta\dot{\mathbf{v}}_k \qquad \text{where} \quad \Delta\dot{\mathbf{v}}_k = \frac{\gamma}{h\beta}\Delta\mathbf{v}_k - \mathbf{d}_k \tag{3.120}$$

$$\ddot{\mathbf{v}}_{k+1} = \ddot{\mathbf{v}}_k + \Delta\ddot{\mathbf{v}}_k \qquad \text{where} \quad \Delta\ddot{\mathbf{v}}_k = \frac{1}{h^2\beta}\Delta\mathbf{v}_k - \mathbf{a}_k \tag{3.121}$$

For a linear system, when the state at time $k + 1$, given by $\Delta\mathbf{u}$, $\Delta\dot{\mathbf{u}}$, and $\Delta\ddot{\mathbf{u}}$ and the associated inertial, damping and internal forces $\mathbf{F}_{k+1}^I$, $\mathbf{F}_{k+1}^D$, and $\mathbf{F}_{k+1}^S$, will satisfy the generalized α equilibrium of (3.100).

3.8.3.1 Equilibrium iterations; Newton iterations

When the residual equation is nonlinear, the increments $\Delta\mathbf{u}$, $\Delta\dot{\mathbf{u}}$ and $\Delta\ddot{\mathbf{u}}$ will in general not satisfy the equilibrium equation (3.100). In order to ensure dynamic equilibrium before advancing to the next time step the dynamic residual seeks to be minimized The linearization of the residual with respect to displacement correction then takes the form

$$\mathbf{R}_\alpha + \frac{\partial\mathbf{R}}{\partial\mathbf{u}}\delta\mathbf{u} = \mathbf{0} \tag{3.122}$$

$$\frac{\partial\mathbf{R}}{\partial\ddot{\mathbf{u}}}\frac{\partial\ddot{\mathbf{u}}}{\partial\mathbf{u}}\delta\mathbf{u} \quad + \quad \frac{\partial\mathbf{R}}{\partial\dot{\mathbf{u}}}\frac{\partial\dot{\mathbf{u}}}{\partial\mathbf{u}}\delta\mathbf{u} \quad + \quad \frac{\partial\mathbf{R}}{\partial\mathbf{u}}\delta\mathbf{u} \quad = -\mathbf{R}_\alpha$$

$$(1 - \alpha_m)\mathbf{M}\frac{1}{h^2\beta}\delta\mathbf{u} + (1 - \alpha_f)\mathbf{C}\frac{\gamma}{h\beta}\delta\mathbf{u} + (1 - \alpha_f)\mathbf{C}\mathbf{u}\delta\mathbf{u} = -\mathbf{R}_\alpha \tag{3.123}$$

We have thus arrived at an iterative equation with the same Newton matrix as in (3.114) and a modified right-hand side:

$$\mathbf{N}\,\delta\mathbf{v}_i = -\mathbf{R}_\alpha \tag{3.124}$$

where

$$\mathbf{N} = \frac{(1-\alpha_m)}{h^2\beta}\mathbf{M} + \frac{(1-\alpha_f)\gamma}{h\beta}\mathbf{C} + (1-\alpha_f)\mathbf{K} \tag{3.125}$$

$$-\mathbf{R}_\alpha = (1-\alpha_f)\mathbf{Q}_{k+1} + \alpha_f\mathbf{Q}_k \tag{3.126}$$

$$-(1-\alpha_f)\mathbf{F}_{k+1}^I - \alpha_f\mathbf{F}_k^I \tag{3.127}$$

$$-(1-\alpha_f)\mathbf{F}_{k+1}^D - \alpha_f\mathbf{F}_k^D \tag{3.128}$$

$$-(1-\alpha_f)\mathbf{F}_{k+1}^S - \alpha_f\mathbf{F}_k^S \tag{3.129}$$

The correction to the displacements, velocities, and accelerations now follows update equations (3.92) and (3.91) when keeping in mind that $\mathbf{a}_k$ and $\mathbf{d}_k$ have already been added to the state at $k+1$

$$\mathbf{v}_{k+1} \mathrel{+}= \delta\mathbf{v}_i \tag{3.130}$$

$$\dot{\mathbf{v}}_{k+1} \mathrel{+}= \frac{\gamma}{h\beta}\delta\mathbf{v}_i \tag{3.131}$$

$$\ddot{\mathbf{v}}_{k+1} \mathrel{+}= \frac{1}{h^2\beta}\delta\mathbf{v}_i \tag{3.132}$$

3.8.3.2   Stability and accuracy

The generalized α algorithm, integrating equation of (3.100), is a second-order accurate algorithm provided that

$$\gamma = \frac{1}{2} - \alpha_m + \alpha_f \tag{3.133}$$

The main purpose of the algorithm is to improve stability and provide additional damping in the high-frequency range (relative to the time step Δt); see Figures 3.8 and 3.9. The generalized α algorithm is unconditionally stable if the following conditions are met:

$$\alpha_m \leq \alpha_f \leq \frac{1}{2} \quad \text{and} \quad \beta \geq \frac{1}{4} + \frac{1}{2}(\alpha_f - \alpha_m) \tag{3.134}$$

According to [9], the principle roots can be expressed as

$$\lambda(\Omega) = A(\Omega) \pm iB(\Omega) \quad \text{where} \quad \Omega = \omega\Delta t \tag{3.135}$$

In order for the high frequency modes (relative to the time step Δt) to be damped out, one wants the roots to be purely real, i.e., $\lim_{\Omega\to\inf} B(\Omega) = 0$. This condition is satisfied if

$$\beta = \frac{1}{2}(1 - \alpha_m + \alpha_f)^2 \tag{3.136}$$

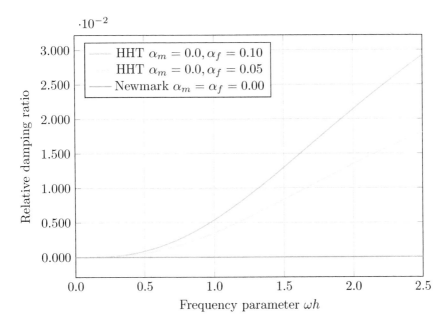

FIGURE 3.8: Numerical damping ratio for the Newmark-integration of the generalized-α algorithms

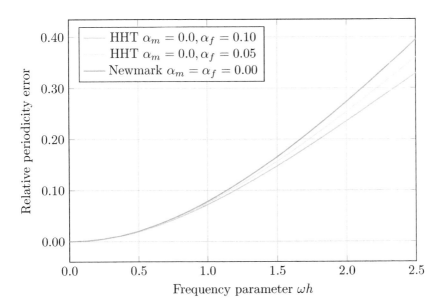

FIGURE 3.9: Relative periodicity error for the Newmark integration of the generalized-α algorithms

4

Framework for Generic Mechanism Modeling

4.1 The Sheth–Uicker (SU) Formulation

The presentation notation used in this section is based on the Bongardt paper [6].

Since the early 1970s the Sheth–Uicker formulation [50] has been the natural choice in our mechanism modeling, but in 2013 this formulation was revisited by Bongardt and the generic features of this formulation were pointed out more clearly. In the precursor book [51] we chose a more straightforward and explicit mechanism formulation, but in order to use a more generic mathematical symbolism, a subset of Bongardt's formulation is used in this presentation.

According to Bongardt a *frame* F is a term used to specify a local coordinate system that may move in space, however, a *pose* $\mathbf{P}$ is a term used to specify a *frame* at a certain time-step.

A *pose* $\mathbf{P}$ describes a rotation and a translation relative to the origin. If a *pose* is associated with a *frame*, it is marked as $\mathbf{P} = \mathbf{P}_F$. In matrix form a *pose* is given as

$$\mathbf{P} = \begin{bmatrix} \mathbf{x} & \mathbf{y} & \mathbf{z} & \mathbf{p} \\ 0 & 0 & 0 & 1 \end{bmatrix} \tag{4.1}$$

where $\mathbf{x}$, $\mathbf{y}$ and $\mathbf{z}$ are the unit vectors of x-, y- and z-direction, respectively, for the *pose*. $\mathbf{p}$ is the position coordinates for the origin of the *pose*. This is also defined in Section 3.1.2.

A set of k frames are denoted $\mathcal{F} = (F_1, F_2, \cdots, F_k)$. A mechanism is composed of a set of n links L and using the same formalism this may be written $\mathcal{L} = (L_1, L_2, \cdots, L_n)$. A mechanism will also have a set of m joints connecting the links denoted as $\mathcal{J} = (J_1, J_2, \cdots, J_m)$. A mechanism could then be denoted as the tuple $\mathcal{M}(\mathcal{L}, \mathcal{J})$.

For the Sheth–Uicker convention there are exactly two frames at each joint in total for a frame set of size $|\mathcal{F}| = 2 * |\mathcal{J}|$. For the SU two-frame convention a mechanism can now be defined as $\mathcal{M}(\mathcal{L}, \mathcal{J}, \mathcal{F})$.

Indicating time-variant symbols by overscore $\overline{(.)}$ and time-invariant symbols by underscore $\underline{(.)}$, a closed kinematic chain could be computed from

$$\mathbf{D}_{(1,1)} = \overline{\mathbf{D}}_1\underline{\mathbf{D}}_{1,2}\overline{\mathbf{D}}_2\underline{\mathbf{D}}_{2,3}\overline{\mathbf{D}}_3\underline{\mathbf{D}}_{3,4}\cdots\overline{\mathbf{D}}_{(n-1)}\underline{\mathbf{D}}_{(n-1),n}\overline{\mathbf{D}}_n\underline{\mathbf{D}}_{n,1} = \mathbf{I} \qquad (4.2)$$

Equation 4.2 is a simplification of the Bongardt formulation and represents the transformations following a closed kinematic chain with n joints starting and ending in the global coordinate system for the mechanism. This represents a transformation following the kinematic chain back to itself, and is thus equal to the unit transformation. In other words, equation 4.2 models time variant transformations over joints and time invariant transformation over links. All frames of the actual kinematic chain are involved.

4.2   Generic Library Format for Mechanisms

There are two ways to model a mechanism, either to model the parts (links) and position them into a mechanism assembly, or to position the joints and extract the link geometry from the positioned joints. To be able to parameterize the overall mechanism, the last mentioned method is easier to handle, especially for closed loop mechanical systems. The entities used to model a mechanism in this way are described in [24]. Mechanisms developed may be stored digitally in a format and structure defined in this section. The proposed format covers the mechanism basics that are necessary for design and simulation. The four-bar mechanism in Figure 4.1 is the reference for the tables in the next section. Only the most basic parameters are included in the tables shown in the next section.

4.2.1   Node positions

A list of 3D points at system level is utilized for defining key positions in a mechanism. The different nodes include: joint positions, connecting points for springs and dampers, point of attack for a force or torque and design points for link geometries. These nodes represent the external nodes of a mechanism. An example of a node position list is shown in Table 4.1. The table indicates that a spring is defined, however, this spring is not shown in Figure 4.1.

4.2.2   Constraints

The mechanism joints are described in a joint list. This list specifies joint type, which links the joint is incident on, the direction, i.e., the orientation of the joint, and the joint's DOFs. An example constraint list is shown in Table 4.2. The numbers in the "Node ref." column refer to the corresponding node in the coordinate list. The "Type" column indicates the joint type, annotated

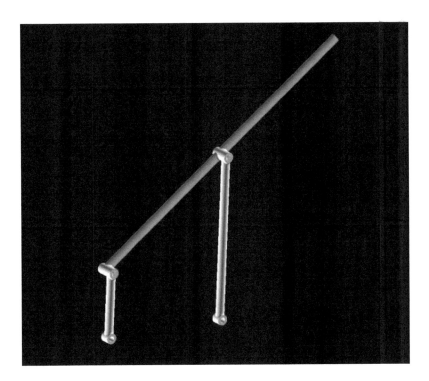

FIGURE 4.1: The four-bar mechanism used as example for input list. Referred to as the Hocken's linkage [35]

.

TABLE 4.1

Example coordinate input list [24].

Index	Label	X-Pos	Y-Pos	Z-Pos
0	"Crank-bearing"	0.0	0.0	0.0
1	"Crank-top"	0.0	0.75	0.0
2	"Rocker-bearing"	1.5	0.0	0.0
3	"Rocker-top"	1.5	1.875	0.0
4	"Coupler-end"	3.0	3.0	0.0
5	"Spring-top"	3.0	3.0	-0.12
6	"Spring-ground"	3.0	0.0	0.0

TABLE 4.2

Example constraints input list [24].

Node ref.	Type	Link-incidence	Joint-dir.	Fixed-DOFs
0	"revolute"	(nil 0)	(0 0 1)	
1	"revolute"	(1 0)	(0 0 1)	
2	"revolute"	(nil 2)	(0 0 1)	
3	"revolute"	(1 2)	(0 0 1)	
4	"free"	(nil 1)	(0 0 1)	

TABLE 4.3

Example link shape input list [24].

Lable	Link	Member	Cross-section	Dimensions
"Crank"	0	0	"circular"	(0.1 0.1)
"Coupler lower"	1	0	"circular"	(0.1 0.1)
"Coupler upper"	1	2	"circular"	(0.1 0.1)
"Rocker"	2	0	"circular"	(0.1 0.1)
"Coupler invis"	1	1	"nil"	

in quotation marks. "Link-incidence" defines the two links that the joint constrains. For example, on the line referring to node 3, (1 2) means that link 1 and link 2 are constrained by a revolute joint. Furthermore, it is also implied that the first element (1) is the solid, male, part of the joint, while the second (2) is the hollow, female, part. One of these values can also be nil. In AML, nil is a placeholder for an empty set, so nil in this context simply means that the given element should not exist. Taking the first input line as an example,(nil 0) means that link 0 should have a female revolute element, located at point 0, and not connected to any male element. A joint having a nil element is a direct indication that it is supposed to be connected to ground.

Free joints have no joint geometry and are initially defined to have six unconstrained DOFs. These DOFs may be specified in the "Fixed-DOFs" column. For instance, (4 5 6) means that the joint is free to translate along the x, y and z direction but cannot rotate about any axis. Analogously, (1 2 3) means that the joint is free to rotate about its x, y and z axis but cannot translate in any direction. Finally, notice that the column "Fixed-DOFs" is left blank. This is because a joint's DOFs are explicitly given by its type. As may be expected the planar four-bar mechanism has four revolute joints in Table 4.2. For the end of the coupler link a free joint is specified, that is a joint with no constraints, and as may be seen none of its degrees of freedom is specified as fixed in column "Fixed-DOFs".

4.2.3   Link shapes

The general appearance of the links is defined in the Link Shape list. Example entries are listed in Table 4.3.

TABLE 4.4
Example link shape input list using default values [24].

Lable	Link	Member	Cross-section	Dimensions
"Coupler invis"	1	1	"nil"	
"Other-links"	default	default	"circular"	(0.1 0.1)

A member is a connection between two joints in a link; refer to Section 4.3. Each link member has specified its type of cross section and its dimensions. The dimensions are defined either as (height width) or (height-start width-start height-end width-end). Height and width refer to the y and z direction, respectively, in the cross section's local coordinate system. The "-start" and "-end" postfixes for the member dimensions defines the same values, only for the start and end cross sections of the link member.

The format for link shapes has two more columns available than shown in Table 4.3: "Nodes" and "Weights". These allow the user to specify control, or design, points for a member. A list of values in the "Nodes" column will refer to the corresponding nodes in coordinate list, and will control the sweep along a member. The sweep follows a NURBS (Non-uniform rational B-spline) curve based on these nodes. The values under "Weights" specify the weighting of each point. If the weight is equal to 1, the curve is simply a B-spline. More details about definition of NURBS are found in Section 4.3.1 and Section 8.2.1 in connection with the suspension system example.

Note that each member of the links is specified in Table 4.3. This does not always have to be the case, as the automatic generation of link geometry is possible. By introducing a "default" value instead of a link and/or a member number, the system can use these values for all the corresponding geometries, as seen in Table 4.4. The first line specifies the coupler link member that should not be generated from the top of crank link to the end of the coupler link. The second line says that all the other link members should be "circular" and with dimension set to (0.1 0.1), that is exactly the same mechanism link as specified in Table 4.3. This functionality allows one to quickly come up with default link geometries for early design stages. For more details about link shapes, refer to Section 4.3.

4.2.4 Springs and dampers

A mechanism might include springs and/or dampers. These are defined in the spring-damper list. An example is illustrated in Table 4.5.

Springs and dampers are treated similarly. To differentiate between the two, the type has to be specified in the "Type" column. To define the start and end points of the springs and dampers, the nodes in the coordinates list are used as references. The links that the spring or damper are connected to has to be defined in the "Links" column. In the last column of Table 4.5 the stiffness or damper coefficients are given, respectively.

TABLE 4.5

Example spring/damper input list [24].

Type	Node-start	Node-end	Links	Stiffnes/Damping
"Spring"	5	6	(1 nil)	100.0
"Damper"	1	2	(0 2)	5.0

TABLE 4.6

Example loads input list [24].

Type	Node	Direction	Magnitude	Link
Torque	0	(0.0 0.0 -1.0)	10.0	0

4.2.5 Loads

For analysis and simulation purposes, loads can also be applied to a mechanism. The loads are defined in the loads list, as seen in Table 4.6.

There are two main types of loads that can be added; forces and torques. As springs and dampers, these are also added to a node as well as defined on a link. One can choose between two types of magnitudes, a constant or a scaling of a ramp function. The constant will instantly apply a force of the given magnitude, while the scale is a simple linear function with the specified magnitude as the slope. For instance, scale30 means a ramp function with slope of 30. There is usually more than one row in a loads list. The actual load definition in Table 4.6 is a torque that will tend to rotate the crank link around its bearing, for instance by an electric motor.

4.3 Default Link and Joint Shapes

All links are considered connections between joints. They can be characterized as a collection of members, where each member connects two joints. A link will have a topological configuration determined by the degree of the link. The shape and dimensions of the members will determine the morphology.

Each link has a number of joints n, attached to it. For a binary link ($n = 2$) there is only one possible connection between the two joints. For a ternary link ($n = 3$) there are three possible connections between the three joints. Calculating all possible connections between n numbers of joints on a link is called a complete graph problem. The complete graph on n joints is denoted by K_n. The number of possible connections is given by $\frac{n}{2}(n - 1)$ [52]; see Figure 4.2.

Each edge in the graph K_n is equivalent to a member on the link connecting two joints. For the link to be connected each joint has to be connected to at least one other joint. Not all connections on a link are necessary; for K_3 and

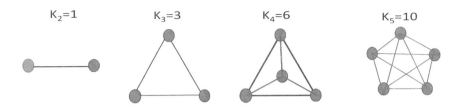

FIGURE 4.2: Number of possible connections between joints [52]

higher the user can select the preferred connections. To deselect a member of a link the "Cross-section" type is marked as `"nil"`, refer to Tables 4.3 and 4.4.

4.3.1 Connection shape

The shape of a member connecting to constraints (joints) is represented using a spline curve. Splines are commonly used in computer graphics for generating and representing curves and surfaces. Splines are defined by interpolating control points to obtain smooth continuous functions. The start and end points of the curve are given by the SU frames in the corresponding joints. Further points in-between are either automatically generated or given by design points in the link shape input list (see Section 4.2.3). This makes it possible to achieve a high degree of flexibility, but still generate a simplified curve when no extra control points are specified.

NURBS (non-uniform rational basis splines) generalize both B-splines and Bézier curves. Like B-splines, they are defined by their order, and a knot vector, and a set of control points determining the shape of the curve, but unlike simple B-splines, the control points each have a weight. When the weight is equal to 1, a NURBS is simply a B-spline.

An additional frame is placed at the start and the end of the connection, with the x–axes tangential to the spline curve. These will be used for positioning the cross section of a member.

4.3.2 Sweep

To make a solid geometry of a member, the splines are used as guides for sweeping cross sections between the constraints, creating a sweep-object. Any

FIGURE 4.3: Sweeps with different varying orientation and different sized cross section [52]

desired cross section can be used, typical types are included and additional types can be added.

A cross section is placed at the start and end of each member. The start and end cross section has a reference to the frames placed at the start and end points. This allows us to create a variational sweep, often seen on different links in mechanisms such as a car suspension.

A few different examples of cross sections are shown in Figure 4.3 where the start and end cross section matches for different orientations and sizes. This works for any new cross-section added.

4.3.3 Surfaces

If combinations of 3 visible connections (members) make a closed loop, a surface is created by default. If the three member splines are connected creating a loop it is possible to create a surface interpolating these splines. The number of possible surfaces is given by $\frac{n}{6}(n-2)(n-1)$, where n is the number of constraints on a link [52]. For this formula to hold no members could be suppressed.

To make sure only members that are displayed will be used to create a surface, their indexes are collected and the combinations are calculated and compared to the possible configurations. The possible surfaces are then created and a surface thickness is added.

4.3.4 Joint geometry

As discussed above the geometry for links is generated by extruding members (connections) between joints and in some cases generating surfaces between members. To complete the link geometry also the geometry for the joints themselves needs to be generated. Each link has references to its joint elements. To model the solid geometry of a joint, the dimensions are determined from the members connected to the joint's pair element. The reason for this is based on the idea that the joints are transferring the forces between links, if the links are sufficiently sized the joints can be expressed as a function of the link's dimensions. The joint elements are chosen to be a factor bigger than the dimensions of the attaching members. An exact functional relationship between link and joint dimensions can be derived.

A so-called main-frame for a joint is positioned and oriented according to the input specifications defined for instance in Table 4.2. What is called a sub-frame is offset along the joint axis according to the joint dimension to allow for space between links. The position of the sub-frame is dependent on joint and element type. The main-frame and the sub-frame for a joint are used for positioning the two pair elements of a joint, what is also called the "male" and the "female" part of the joint, respectively.

Dimensions of a joint are found by looping through all members connected to a joint element, if they are currently displayed. It checks if the joint is connected to the start or end cross section, and collects the maximum height and width of the members connected. The maximum width and height of the local and the paired element are then used as parameters for determining the dimensions of the joint and will also be used to determine the placement of the sub-frames as discussed above.

4.3.5 Assembly

Since the joint elements in reality are integrated parts of the link, they are assembled together with the members and the surfaces. This is done by collecting the parts of a joint element which are union or difference into separate lists.

Each link then collects the objects from its connected elements and assembles them using a union-object. At last one difference-object is created. This is to make sure the hollow parts of joint elements are subtracted from the link where necessary. Sharp edges between the merged geometries of links members and link joints are blended by default to allow for high quality FE mesh to be generated for the link geometry. For each link a separate geometric model is generated and is meshed automatically as input for simulation.

4.4 Extended Modeling of Link Shapes

The aim of the parameterized default link shape algorithm described in Section 4.3 is to produce link geometries sufficiently detailed for analysis using the library format from Section 4.2. An interactive interface for editing the library format is available to tune the details of the link geometry to a level of detail required for analysis and simulation; see Chapter 7.

In many cases, the geometric details used for the simulation may be detailed enough for production of the links as well. If further detailing for production is necessary, the parameters in the library format could be used for further modification beyond the simulation geometries or the links could be exported to a CAD system for further detailing to prepare for the production phase. The application has tools for generating STEP files of the link geometries for transfer of the link geometries to any CAD system with STEP format input. Any further detailing and styling for production may then be done in the chosen CAD system.

4.5 Automated Generation of Simulation Input

The simulation tool used here is the FEDEM simulation program supported by FEDEM Technology AS [3]. This simulation tool is used for dynamic analysis of mechanisms. Flexibility of the mechanism links is based on FE modeling. Simulation input includes a separate file for each link defining the finite element mesh and other FE data such as material properties. In addition, a file with system data for the mechanism is needed. This file will refer to the files with FE data for the different links. The system input file also includes link topology for the mechanism, details for joints, springs and dampers, possible control elements, loading, and control data for the simulation.

4.5.1 Finite element data for mechanism links

The mechanism model for a simulation is generated in a KBE application implemented in the KBE language AML [60]. The AML model of the mechanism generates solid geometry for each link including joint geometry of the mechanism and generates FE meshes for the different links automatically using 4 node tetrahedron finite element. The NASTRAN Bulk Data Format (BDF) is used as FE input for the different links. Typical data in each file with link data includes coordinates for nodal points in the mesh, mesh topology, i.e., how the different finite elements are connected to the nodal points. Material data is also included in these files. The link files in BDF format is generated automatically by a KBE application using AML. These files are processed by the simulation code to produce a FE substructure with stiffness, damping and mass matrices. These substructure matrices are condensed to super element matrices using the component mode synthesis (CMS) method [51], and shown in Section 3.4.3. During each time step of the simulation these super element matrices are added to the mechanism system matrix as input to the time integration.

4.5.2 Mechanism system data

To put together the mechanism for simulation a separate file is used. This file has references to the FE data for the different links; refer to previous subsection. As mentioned above the super element matrices for each link are transformed to the actual direction for the different system nodes and added into the system matrix at mechanism level. In addition to the link input from BDF files details for the different mechanism joints are generated: What links are involved in the different joint, joint type, joint direction, etc. There may be springs and dampers between links and/or in joints. Springs and dampers may be linear or nonlinear and separate functions are generated to describe any nonlinearity. Forces and torques may be specified on links and/or in joints. If control functionality is used control elements and control connectivity are specified. Control will often require virtual sensors for measurements in the simulation model and torques and forces may have their values generated by the control system. The system file may also specify eigenvalue analysis of the mechanism system at certain time intervals.

The system file will also specify control parameters for the execution of the simulation like integration time step, tolerances for convergence, parameters to be plotted during simulation, if strain and stress retracking should be specified and for which link and what time steps. A lot of other details are also specified in the simulation system input; for more details refer to the FEDEM user manual [3]. Setting up the detailed inputs to the FEDEM system file is both very time consuming and also very prone to human errors. Generating this file automatic is very advantageous both with respect to time and quality when doing simulation. This will also make it possible to use optimization techniques to optimize the actual product both with respect to integrity and functionality, using repeated simulations and searching for solutions with improved performance.

4.6   Example Mechanism with Different Modeling Elements

The key modeling elements in our pilot KBE tool for automated mechanism design is demonstrated in an extended version of the four-bar mechanism shown earlier in this chapter; see Figure 4.1. Our demonstration mechanism is shown in Figure 4.4. This figure also shows the graphic environment for KBE modeling in the AML framework [60].

The input data to generate this mechanism is shown in Table 4.7. This figure is a compilation of the type of data presented earlier in this chapter as a set of tables. To demonstrate functionality of the modeling tool, the coupler link is composed of two links, a compound link with four different joint types: a ball joint on top of the crank, a revolute joint on top of the rocker, a fixed joint (also call a rigid joint in the FEDEM simulation tool [3]) to rigidly connect the tip of the coupler modeled here as a separate link, and at last what is called a free joint, that is a joint with no constraints as default and different user defined constraints could be added at a later stage in the FEDEM simulation tool.

The input data also include example definitions of an axial damper, an axial spring, and also the loading of the mechanism as a torque driving the crank of the mechanism; see illustrations in Figure 4.4. The axial damper is put between the rocker bearing and the crank top to limit the crank speed during simulation. With no damping and the crank bearing being loaded with a constant torque, the rotational speed of the crank would increase linearly (and indefinitely). An axial spring is put between the rocker bearing and the coupler just to demonstrate axial spring modeling. A torque with direction along the negative z axes is modeled in the crank bearing as mentioned above. The torque is illustrated as a double arrow. See the fourth and fifth section of the input data in Table 4.7 regarding the details of damper, spring and load modeling.

4.6.1   Example joint definition input

In the second section of the input data, all joints in this mechanism are defined. In the line referring to point 1, the ball joint between the crank link and the coupler link is defined. Below "Link-incidence" we have specified (1 0) that refers to the coupler link and the crank link, respectively. In our terminology this means that the "male" part of this ball joint sits on the coupler link (referred to first) and the "female" part of this joint sits on the crank link. Playing around with which link is referred to first will have consequences for both how the link and joint geometries are generated. The mechanism defined here is planar and is defined in the xy-plane. Below "Joint-direction" we have specified (0 0 1) for this joint, defining that the joint direction should be parallel to the positive z-axis, that is the joint direction has no component along the x- or y axis and a unit component along the z axis. The line in the same input section referring to point 3 defines the revolute joint between the coupler link (1) and the rocker link (2). Also here the "male" part of this revolute joint sits on the coupler link and hence the "female" part sits on the rocker link that is link 2. The line in this joint section for our mechanism referring to point 5 is a joint called "fixed" (also named rigid) and is used to connect the coupler link rigidly to the link (3) used to define the end of the coupler link, defined here as a separate link for demonstration purposes. The line referring to point 6 defines a free joint that is a joint with no constraints as default. The free joint is defined between the ground link (nil) and the coupler link (1).

The remaining joint definitions in the joint section of the input data (see Table 4.7) is the line referring to point 0 that is a revolute joint with the "male" part sitting on the ground link (nil) and the "female" part sitting on the crank link (0). The line referring to point 2 is also a revolute joint between the ground link (nil) and the rocker link (2). And at last the line referring to point 4 defines a second free joint between the end of the link (3) being the rigid extension of the coupler link (1). The coupler curve for this straight line generator mechanism is generated for the same position as for this last free joint.

The points referred to above in the joint section of the input data are referring to the first section of the input data; see Table 4.7. This section defines the x, y and z coordinates for positioning the different joints and other entities. As described above the joint section (section 2) of the input data defines the joint topology for a mechanism while section 1 are the primary design variables for the mechanism geometry.

4.6.2 Example link shape definition input

Taking a look at the modified coupler link (1), the four joints are referring to node 1, 3, 5 and 6, and with four joints we have by default 6 members as connections between these joints. The six members are (1 3), (1 5), (1 6), (3 5), (3 6) and (5 6) named member 0 through 5, respectively. We choose a rectangular cross section for member 0 connection (1 3) which in line two is named "coupler-lower" in section three of the input data; see Table 4.7. Members 1 and 2 are not needed for this demo mechanism and are suppressed by giving a cross-section value of "nil"; see lines 3 and 4 of the input data in section 3. We have chosen to give member 3 (3 5) named "coupler-upper" "circular" cross-section; see line 5 of the shape section of the input data. The members 4 (3 6) and 5 (5 6) to the connection for the spring are given cross-section type "rectangular-tube". As members 3, 4 and 5 make a closed loop, a surface is generated by default between these members. We have chosen to give the rectangular members a height and a width of 0.1, and the circular members are also given a diameter of 0.1.

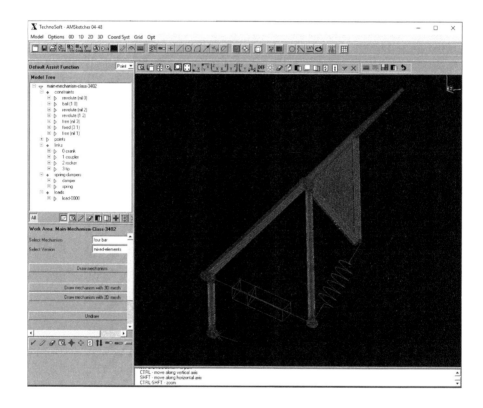

FIGURE 4.4: Demonstration four-bar mechanism

TABLE 4.7

Input data for the demonstration four-bar mechanism.

Index	Name	X-pos	Y-pos	Z-pos
0	"Crank-bearing"	0.0	0.0	0.0
1	"Crank-top"	0.0	0.75	0.0
2	"Rocker-bearing"	1.5	0.0	0.0
3	"Rocker-top"	1.5	1.875	0.0
4	"Coupler-end"	3.0	3.0	0.0
5	"Coupler-split"	2.25	2.4375	-0.12
6	"End-down"	2.25	1.0	0.0

Point	Type	Link-incidence	Joint-dir
0	"revolute"	(nil 0)	(0 0 1)
1	"ball"	(1 0)	(0 0 1)
2	"revolute"	(nil 2)	(0 0 1)
3	"revolute"	(1 2)	(0 0 1)
4	"free"	(nil 3)	(0 0 1)
5	"fixed"	(3 1)	(0 0 1)
6	"free"	(nil 1)	(0 0 1)

Name	Link	Member	Cross-section	Dimensions
"Crank"	0	0	"circular"	(0.1 0.1)
"Coupler lower"	1	0	"rectangular"	(0.1 0.1)
"Coupler invis"	1	1	"nil"	
"Coupler invis"	1	2	"nil"	
"Coupler upper"	1	3	"circular"	(0.1 0.1)
"Coupler upper"	1	4	"rectangular-tube"	(0.1 0.1)
"Coupler upper"	1	5	"rectangular-tube"	(0.1 0.1)
"Rocker"	2	0	"circular"	(0.1 0.1)
"tip"	3	0	"circular-tube"	(0.05 0.05 0.1 0.1)

Type	Point-from	Point-to	Incident-links	Stiffness/Damping
"Damper"	1	2	(0 2)	5.0
"Spring"	2	6	(2 1)	10.0

Type	Point	Direction	Magnitude	Loaded-link
Torque	0	(0.0 0.0 -1.0)	50.0	0

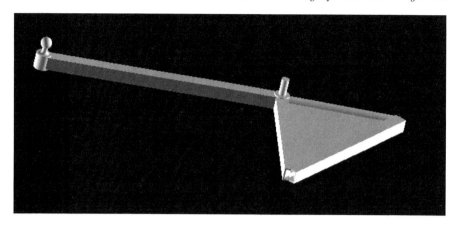

FIGURE 4.5: Coupler link geometry

For the four joints implemented and tested in the pilot KBE application as demonstrated on this coupler link, the ball joint and the revolute joint generate what we call a "male" or "female" geometry for the joint; see Figures 4.4 and 4.5. The free joint and the fixed joint have not defined a joint geometry. In the center of each joint is a coordinate system named main-frame, which is in the position defined in the input data and referred to by the joint. For point to point joints like the ball and revolute joints, the master and slave triads as defined in FEDEM, will both be in this position. See also equations (3.52) and (3.53) in Section 3.5.

Most joints have also defined a sub-frame that is displaced in negative direction from the main-frame along the joint axis. This sub-frame is the connecting point for members into the joint; refer to the description earlier in this chapter. The largest dimension of a member coming into the joint defines the joint dimension by multiplying this largest dimension by the factor 1.2. This factor is also used to displace the sub-frame from the main-frame. The ball, revolute and free joint automatically displace the sub-frame this distance along the joint axis. However, we see no reason to displace the sub-frame from the main-frame for the fixed joint. This is the reason that we manually displaced node 5 in the input data 0.12 along the negative joint axis; that is the dimension of the connecting member multiplied by the factor 1.2. By making this displacement of node 5 this coupler link will be parallel to the xy plane.

Taking away the crank link, the rocker link, and the tip of the coupler, the modified coupler link is retained; see Figure 4.5. In this figure we see the "male" part of the ball joint to the left as a ball with a connection to a cylinder containing the joints sub-frame, the "male" part of the revolute joint in the middle as a cylinder connected to a larger cylinder containing its sub-frame. In the closest point, we see the free joint where the spring is connected. The rectangular tube members ending in this joint are visible here. The free joint has no defined joint geometry. To the right in the figure we see the position of the fixed joint that does not have a defined joint geometry either. Between members and joint geometries there may be sharp edges and default blending of these sharp edges is implemented to improve quality of the FE mesh. This will be discussed further when FE meshing is presented.

The remaining shape entries in section 3 of the input data in Table 4.7 are the geometric entries of the crank, rocker and tip link. The crank link (0) is defined in line 1 to have one member (0) with circular cross-section and diameter 0.1. In the lower end it has a "female" revolute joint geometry and at the upper end it has a "female" ball joint element; refer to Figure 4.4. In line 8 of the input data the rocker link (2) is defined in a very similar way, but having a "female" revolute joint element in both ends. The tip link (3) is defined in line 9 of the input data. This link has no joint geometry included as it is connected to the coupler link with a fixed joint and the end is connected to the ground link with a free joint and none of these two joints has a geometry defined. The tip link (3) has a varying cross-section with diameter 0.1 close to the coupler and diameter 0.05 at the outer end (the direction of this link is from the end to the coupler link).

4.6.3   Details about FE meshing of links and joints

All links of the mechanism shown in Figure 4.4 have a solid tetrahedral FE mesh generated. FE mesh details for the ball joint at the top of the crank link are shown in Figure 4.6. The upper left view in this figure shows the FE mesh for the "male" and the "female" part of the joint put together. The blending (rounding) of the sharp edges is easy to see in this view including the refined FE mesh in the blended area. The sphere at the top of the crank link is hollow to make room for the ball of the "male" part. The upper right view in this figure shows the "male" part of this ball joint with a sphere representing the ball joint surface. This ball has a connection to a somewhat larger cylinder which is connected to the link member. The joint sub-frame is located in the middle of this larger cylinder as a connecting point for link members entering into this joint. Looking closely at the upper right view, you can see that the nodes on the ball surface have light gray dots which represent the dependent nodes for the joints RBE2 on the "male" part of the joint. Details about RBE2s, in general called multi point constraints (MPC), are discussed in Section 3.7. The light gray dots representing the dependent nodes of the RBE2 are shown separately and somewhat enlarged in the view at the bottom of the figure. Looking very closely you may see a dark gray point in the center of the ball that is the independent node that represents the connecting point to the coupler link.

The inner surface cavity space in the sphere at the top of the crank link has also dependent nodes on the FE mesh nodes for another RBE2 connection for a "female" part of the joint. This RBE2 connection also has an independent

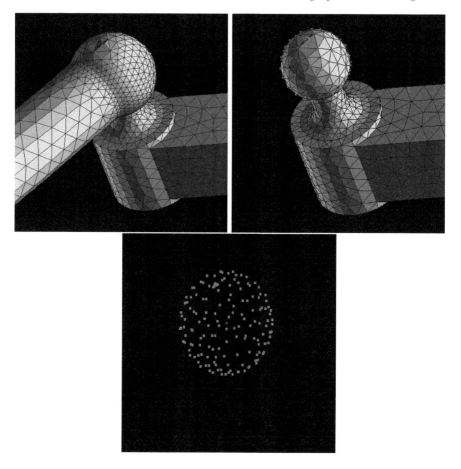

FIGURE 4.6: Mesh details for ball joint

node in the center for the connection between the crank and the coupler link in the simulations. The RBE2 connection on the crank link is not shown in the figure.

FE mesh details for the revolute joint at the top of the rocker link are shown in Figure 4.7. The upper left view in this figure shows the FE mesh for the "male" and the "female" part of the joint put together. The blending (rounding) of the sharp edges is easy to see in this view including the refined FE mesh in the blended area. The cylinder at the top of the coupler link has an inner cylindric opening that represents the joint surface for the "female" part with space for the cylinder of the "male" part. The upper right view in this figure shows the "male" part of this revolute joint with a cylinder representing the revolute joint surface. This cylinder is connected to a somewhat larger cylinder that is connected to the link member. The joint sub-frame is located

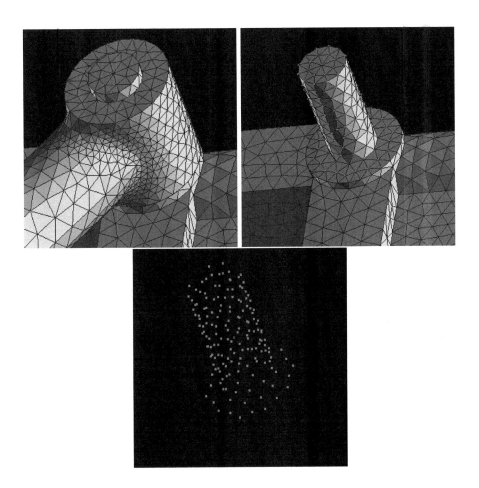

FIGURE 4.7: Mesh details for revolute joint

in the middle of this larger cylinder as a connecting point for link members entering into this joint. Looking closely at the upper right view you can see that the mesh nodes on the cylinder surface have light gray dots which represent the dependent nodes for the joints RBE2 on the "male" part of the joint. Details about RBE2s, in general called multipoint constraints, are discussed in Section 3.7. The light gray dots representing the dependent nodes of the RBE2 are shown separately and somewhat enlarged in the view at the bottom of this figure. Looking closely, you may see a dark gray point in the center of the cylinder that is the independent node that represents the connecting point to the coupler link in the simulations.

The inner cylinder surface of the cylinder at the top of the rocker link also has dependent nodes on the FE mesh nodes for another RBE2 connection for a "female" part of the joint. This RBE2 connection also has an independent node in the center for the connection between rocker and coupler link. The RBE2 connection on the rocker link is not shown in the figure.

4.6.4   Importing demonstration example into FEDEM

The demonstration mechanism is also imported into the FEDEM [3] simulation environment; see Figure 4.8. The graphic environment for modeling and simulating in FEDEM is also shown in this figure. You can recognize the same elements as in the KBE model. Be aware that the links are numbered in the KBE environment from 0 (zero), but in FEDEM they are numbered from 1 (one).

In Figure 4.9 four snapshots from animation of the FEDEM simulation are shown. The view up to the left is passing the start position after one rotation of the crank link. Up to the right the crank link has rotated 90 degrees from the start position. In the view below to the left the crank has rotated 180 degrees from the start position and in the view below to the right the crank has rotated 270 degrees from the start position. In the FEDEM environment revolute joints are modeled as circles and ball joints are modeled as spheres. Free and fixed joints are modeled as coordinate systems in FEDEM.

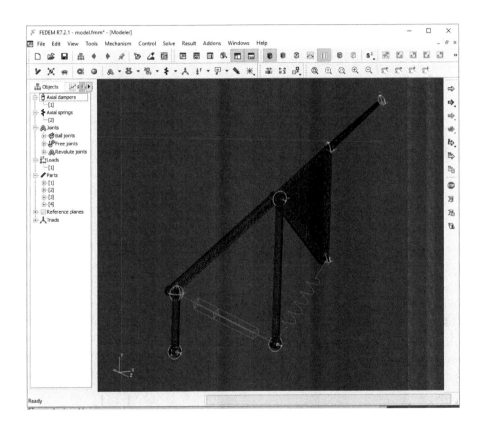

FIGURE 4.8: Demonstration four-bar mechanism in FEDEM

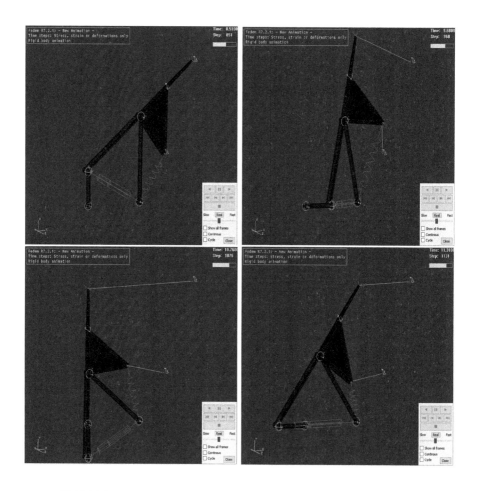

FIGURE 4.9: Snapshots from simulation animation in FEDEM

5

Design Optimization

This chapter is to a large extent based on the work of Sigurd Trier [61] and Bernhard Specht [57, 56]. Jorgen Nocedal's book [34] has also been a resource.

5.1 Formulation of Optimization Problems

Formulation of a design optimization problem involves transcribing a verbal description of the problem into a well-defined mathematical statement in terms of design variables, objective function and constraints. In the following, each of these will be described.

5.1.1 Design variables

Let the *design variables* be a set of independent parameters describing the design of a flexible multibody system. Once we assign numerical values to the design variables, we have a design of the system. By changing the value of one or more design variables we obtain alternate designs. The design variables can include (i) dimensions; (ii) coordinates of joints and attachment points for springs and dampers; (iii) coefficients of springs and dampers; or (iv) any other parameter controlling the design of the system. A design variable can be classified as a continuous or discrete variable. A continuous design variable can be assigned any value, whereas a discrete design variable must take a value from a given set of values. For example, a plate thickness must be selected among those available commercially. A design variable may even represent a function, such as a shape function describing the variation of a cross section dimension over the length of a beam. In this case, the optimization problem is to find the optimal function instead of an optimal value. We do not consider discrete design variables or design variables that represent a function, thus, we assume that the design variables are continuous variables. The design variables x_i are represented by an m-vector $\mathbf{x} = [x_1, x_2, \ldots, x_m]^T$, where m is the number of design variables.

5.1.2   Objective function

The *objective function* is a measure of design quality or performance of the system. To be able to improve the design, the objective function must be a function of the design variables **x**. The objective function will be minimized or maximized, and some typical formulations are to minimize weight, minimize maximum stress, and maximize the fundamental frequency of a system. In the latter two cases, the objective function is an implicit function of the design variables as some kind of analysis of the system must be performed to calculate the responses and thereby evaluate the objective function. In some situations, however, it is impossible to identify a single measure of design quality or performance. In these cases, we may select two or more objective functions to be minimized at the same time. Optimization with more than one objective function is known as multicriteria or multiobjective optimization, and is briefly presented in Section 5.4. In this section we consider an objective function that represents a single measure of design and performance, and we let it be represented by the scalar function $f = f(\mathbf{x})$. Without loss of generality, we may assume that the objective function f is always minimized, since the problem of maximizing f is equal to that of minimizing $-f$.

5.1.3   Constraints

All requirements placed on a design that define acceptable values of the design variables are called *constraints*. They are expressed as functions which must be either within specified limits or satisfy a given limit. Thus, we distinguish between equality and inequality constraints. The constraints must be explicit or implicit functions of the design variables. The given lower and upper bounds on a design variable or the condition that one design variable must be greater than another are examples of such explicit functions. On the other hand, constraints on performance of the system will usually be implicit functions of the design variables as some kind of analysis, such as a structural or mechanical analysis, must be performed to evaluate these constraints. Examples of constraints on performance are that (i) calculated stresses must not exceed allowable stress of the material to prevent material failure; (ii) natural frequency must be higher than the operating frequency to avoid resonance in the system; and (iii) displacements, velocities, or accelerations must not exceed specified limits to satisfy certain performance criteria. We formulate the constraints on a standard form such that inequality constraints must be less than zero and equality constraints must be equal to zero. Let the vectors $\mathbf{g}(\mathbf{x}) = [g_1, g_2, \ldots, g_p]^T$ and $\mathbf{h}(\mathbf{x}) = [h_1, h_2, \ldots, h_q]^T$ represent the p inequality and q equality constraints, respectively.

5.1.4   A standard problem

A design optimization problem can be given as follows: Find the values of the design variables $\mathbf{x}$ which

$$
\begin{array}{ll}
\text{minimize} & f(\mathbf{x}) \\
\text{subject to} & \mathbf{h}(\mathbf{x}) = \mathbf{0} \\
\text{and} & \mathbf{g}(\mathbf{x}) \leq \mathbf{0}
\end{array}
\tag{5.1}
$$

where $\mathbf{g}$ and $\mathbf{h}$ are vectors of inequality and equality constraint functions, and f is the objective function. The lower and upper bounds on the design variables are included in the inequality constraints. The functions in equation 5.1 are assumed to be smooth functions, so they are twice-continuously differentiable with respect to the design variables.

5.1.5   Fundamental concepts

This section presents some fundamental terms and concepts related to optimization and algorithms, such as minimum points, optimality conditions, convex problems and convergence properties.

The optimization problem in equation 5.1 is to find a point in the feasible set, which gives a minimum value to the objective function. A *feasible design* or *feasible point* $\mathbf{x}$ satisfies all constraints, and the *feasible set* is simply the collection of all feasible points. A point $\mathbf{x}^*$ is called a *local minimum point* for the problem in equation 5.1 if $f(\mathbf{x}^*) \leq f(\mathbf{x})$ for all feasible points $\mathbf{x}$ within a small distance of the point $\mathbf{x}^*$. A point $\mathbf{x}^*$ is called a *global minimum point* for problem in equation 5.1 if $f(\mathbf{x}^*) \leq f(\mathbf{x})$ for all points $\mathbf{x}$ in the feasible set.

In plain English this says that finding a minimum point for the objective function does not necessarily represent the optimal solution for the problem, it could be just a local solution and further search could be needed to find the best solution or what we would call a global solution. The global solution does not necessarily represent a mathematical minimum of the objective function; it may be on a slope of the objective function at a constraint border which is the lowest minimum point of the objective function within the legal values for the design variables $\mathbf{x}$.

The conditions that must hold at a minimum point of an optimization problem are called optimality conditions. For local minimum points, these conditions can be established by assuming that we are at an optimum point and then examining the functions and their derivatives at the point. Since there are no optimality conditions to characterize global minimum points, we can only expect to find local minimum points. Global minimum points can, as a rule, only be found if the optimization problem is convex. Convex functions and convex problems are explained below.

The optimality conditions can be divided into necessary and sufficient conditions. The necessary conditions must be satisfied at any minimum point, but there may be other points than minimum points which satisfy them. Thus, violations of these conditions are often used to show that a given point is not a minimum. On the contrary, satisfaction of the sufficient conditions guarantees that a point is indeed a minimum point.

The optimality conditions for a smooth function of $\mathbf{x}$ are expressed in terms of the gradient and the *Hessian matrix* of the function $f(\mathbf{x})$, where the gradient with respect to $\mathbf{x}$ is the vector of first-order partial derivatives, and the Hessian matrix with respect to $\mathbf{x}$ is the matrix of second-order partial derivatives. In the case of optimization problems without any constraints, the optimality conditions are simply extensions of the well-known derivative conditions for a function of a single variable that hold at a minimum point. Thus, the first-order necessary conditions state that at a local minimum point of a smooth function, the gradient of the function vanishes. Further, the second-order necessary conditions state that at a local minimum point of a smooth function, the gradient of the function vanishes and the Hessian is *positive semidefinite*. Finally, the second-order sufficient conditions state that if at a point the gradient vanishes and the Hessian is *positive definite*, then that point is indeed a local minimum point. A matrix with positive eigenvalues is positive definite, and a matrix with non-negative eigenvalues is positive semidefinite.

Let us return to convex functions and convex problems. A function is *convex* if the Hessian matrix is positive semidefinite. In other words, a function is convex if the line joining two points on its graph lies nowhere below the graph. If the Hessian is positive definite then the function is a strictly convex function. The optimization problem in equation 5.1 is said to be a *convex problem* if f is a convex function, the inequality constraints are convex functions and the equality constraints are linear functions. If, in addition, f is a strictly convex function, then the problem is a strictly convex problem. Convex optimization problems are attractive since they guarantee that any minimum point is a global minimum point. Further, for strictly convex problems there is exactly one local minimum point, and this point is also the unique global minimum point.

We end this section by some remarks about algorithms used in mathematical programming methods. Algorithms which are regarded as acceptable have the property of being iterative descent algorithms. An iterative algorithm generates a series of points, each point being calculated on the basis of the points preceding it. Thus, given a starting point $\mathbf{x}(0)$, the algorithm will generate a sequence of points $\mathbf{x}(1), \mathbf{x}(2), \ldots$ hopefully converging to a minimum point $\mathbf{x}^*$. For a descent algorithm the value of some descent function will decrease at every iteration, unless a minimum point is reached. In the case of unconstrained optimization, the descent function is usually the objective function itself.

If the algorithm is guaranteed to generate a sequence of points converging to a minimum point for arbitrary starting points, then the algorithm is said to be *globally convergent*. Not all algorithms have this desirable property, but it is often possible to modify such algorithms so as to guarantee global convergence. One of the conditions to ensure global convergence is that the algorithm has a descent function. On the other hand, local convergence properties are related to how fast the algorithm converges to a minimum point. They can be used to determine the relative advantage of one algorithm to another.

5.2 Optimization Methods

In this section we present some mathematical programming methods for unconstrained and constrained optimization. These methods are not capable of making a distinction between local and global minimum points, and will accept any local minimum as the solution, so the mathematical programming methods are sometimes called local methods. Global optimization methods are also briefly mentioned.

5.2.1 Methods for unconstrained optimization

Mathematical programming methods to solve unconstrained optimization problems include the method of steepest descent, Newton's method, quasi-Newton methods and conjugate gradient methods. These methods are line search methods, so first we find a search direction, and then we perform a line search in that direction to minimize the objective function. The major difference between these methods is how the search direction is calculated. With small modifications, all these methods are globally convergent.

The search direction in the method of steepest descent is the negative gradient of the objective function, as the objective function decreases most rapidly in this direction, at least locally. *Newton's method* uses second-order derivatives to approximate the objective function by a quadratic function, and minimizes this approximate function exactly to obtain the search direction. This means that the search direction is found by solving a linear system of equations at each iteration. *Quasi-Newton methods* replace the second-order derivatives required in Newton's method by approximations based on first-order derivatives at the current and previous iterations, and thereby reduce the computational cost.

Conjugate gradient methods can be regarded as being somewhat between the method of steepest descent and Newton's method. It is motivated by the desire to accelerate the rather slow convergence associated with the method of steepest descent while avoiding the evaluation of the second-order derivatives and the solution of a linear system of equations at each iteration as required by Newton's method. The search direction is found by adding a scaled search direction used in the previous iteration to the current steepest descent direction. Conjugate gradient methods are considered among the best general purpose methods for unconstrained optimization [28].

5.2.2   Methods for constrained optimization

This section gives a brief introduction to a few mathematical programming methods to solve constrained optimization problems. Unless stated otherwise, these methods are line search methods, and with small modifications, they are globally convergent.

Transformation methods are procedures for approximating constrained optimization problems by a sequence of unconstrained problems. These methods [28] include penalty and barrier function methods as well as multiplier methods. Penalty function methods work by adding to the objective function a term that prescribes a high cost for violation of the constraints. A popular penalty function is the quadratic penalty function. The penalty function methods are sometimes called exterior methods because they iterate through the infeasible region, where the infeasible region is the collection of points that violate at least one constraint. Barrier function methods are applicable to inequality constrained problems, and work by adding to the objective function a term that favors points interior to the feasible region over those near the boundary, and the barrier methods are therefore also known as interior point methods. This means that a large barrier is constructed around the feasible region so given a feasible start point, the barrier function methods will create a sequence of feasible points because the iterative process cannot cross the huge barrier. A sequence of feasible designs represents an attractive feature from an engineering point of view, since the designer may stop the iterative optimization process at any stage, and get an acceptable design which is better than the initial design. In multiplier methods or augmented Lagrangian methods, two terms are added to the objective function, both a quadratic penalty function and a term involving so-called Lagrange multipliers.

In *sequential linear programming methods* the objective and constraint functions are approximated by linear functions at each iteration. The corresponding subproblem is convex and can be solved without any line search procedure. So-called move limits and additional safeguards must be imposed to ensure global convergence. In *sequential quadratic programming* (SQP) *methods*, the objective function is approximated by a quadratic function, whereas the constraint functions are approximated by linear functions at the current iteration point. The second-order derivatives of the constraint functions are, however, not neglected; these derivatives are transformed and added to the objective function. To avoid the expensive calculation of second-order derivatives, they are approximated by first-order derivatives at the current and previous iterations, in the same way quasi-Newton methods do for the objective function in unconstrained optimization. So we can say that SQP methods in constrained optimization corresponds to quasi-Newton methods in unconstrained optimization.

The convex approximation methods, such as the *method of moving asymptotes* [59], are a subclass of mathematical programming methods. They were developed to solve inequality constrained problems in structural optimization, motivated by a desire to reduce the number of structural analyses and obtain a sequence of feasible designs. In these methods, the original problem is approximated by a sequence of explicit and convex subproblems which can be solved without a line search procedure. The major disadvantage is that these methods are, in general, not globally convergent, so they tend to be less reliable than the methods presented above.

The computational cost in structural optimization is determined by the number of structural analyses since we must perform one structural analysis at each iteration to evaluate the objective and constraint functions, and the cost of each structural analysis is much more than the cost within the optimizer itself. When the gradients of the objective and constraint functions with respect to the m design variables are approximated by finite differences, we need $m + 1$ structural analyses at each iteration just to find the search direction. The cost of adding a few more structural analyses in the line search procedure at each iteration to ensure global convergence seems to be a fair price to pay.

5.2.3 Global optimization methods

The optimization methods described so far are local methods in the sense that they converge to a local minimum point. Global methods have a much higher ambition, as they are determined to find a global minimum point, or at least an approximation to a global minimum point. Typically, they evaluate the function values at many points distributed over the entire feasible region, and this requires many structural analyses.

A class of optimization methods called guided random search techniques (GRST) [54] is planned for searching the entire feasible design space for an optimal solution. The best known technique in this class is the method called genetic algorithms (GA). These methods are not referred to often in connection with structural design, however, parallel computing could make these methods effective also in the field of structural design, for instance in combination with more classical optimization methods. These methods are not discussed further here. More details may be found in [54].

5.3 Multidisciplinary Design Optimization

Response and design optimization based on dynamic simulation of flexible multibody systems (FMBS), including control systems, clearly fall within the category of *multidisciplinary design optimization* (MDO) as defined in Sobieski et al. [54], however, the size of the problem and the number of design variables **x** are in many cases quite small. Besides FMBS including the mechanism dynamics, structural dynamics and often control simulation are very multidisciplinary, but also very concurrent and coupled and don't lend themselves easily to computational parallelization. A unified approach to optimization also works well because the designers need to be quite well acquainted with all fields involved in the overall FMBS simulation, at least some of the design team. Based on these assumptions we limit our approach to what is named single-level design optimizations (S-LDO) according to [54]. This is mainly what is presented up to now in this chapter. This is also the simplest and most straightforward approach that has been the state-of-the-art design optimization technique for many years. According to [54] S-LDO is proposed as the initial step of design optimization with a reduced set of design variables also for very complex multidisciplinary design simulations. Rapid advances in massively concurrent computing makes it likely that the role of S-LDO will be extended in the future [53].

Because the simulation models we are discussing in this text are generated automatically, the team manipulating the simulation parameters needs to be quite acquainted with all technologies involved, and will benefit from a unified optimization approach. For very large and complex simulation models, the S-LDO could have limitations that could require a more modular approach to design optimization. Also in the event that the method of automated model generation for FMBS presented in this text is included in an overall simulation of a larger system the FMBS optimization could be a module in a larger optimization scheme; refer to Sobieski et al., for more details about this [54].

5.4 Multiobjective Optimization

As the name indicates, multiobjective optimization means that instead of optimizing according to one objective function $f(\mathbf{x})$ you should optimize with respect to two or more objective functions, that is a vector of objective functions

$$\mathbf{f}(\mathbf{x}) = [f_1(\mathbf{x}), f_2(\mathbf{x}), \ldots, f_q(\mathbf{x})]^T \qquad (5.2)$$

In FMBS you could consider optimizing with respect to how well the system achieves the function requirements, for instance how well a point on a link follows a prescribed path and at the same time minimizing the weight of the system. A third objective could be to minimize the cost for producing the system. How to balance between the different objectives is a challenge that requires interaction with the designer and the optimization tool.

A nineteenth-century Italian mathematician, Vilfredo Pareto, established a mathematical foundation for multiobjective optimization. As an illustration see the objective functions f_1 and f_2 plotted against a single design variable x in Figure 5.1. The minimum point for f_1 and f_2, respectively, A and B define the interval P between points Q and R on the x-axis. This is the only interval where f_1 increase and f_2 decrease. Pareto says that each point in

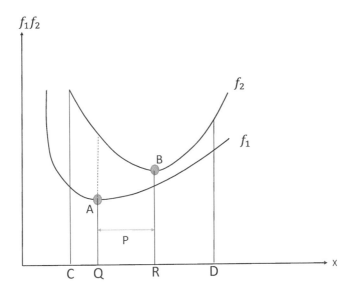

FIGURE 5.1: Pareto optimization [54]

the P interval is a candidate for a multiobjective optimization result. The accepted Pareto-minimum definition generalized for n objectives: A multiobjective Pareto-minimum point in the **x**-space is one for which it is impossible to depart without making at least one of the n objectives worse [54].

One straightforward procedure to search for a multiobjective optimization solution is to make one single objective function as a weighted sum of all the objective functions and to use a search procedure for a single objective optimization problem (equation 5.3)

$$\text{minimize} \quad f(\mathbf{x}) = \sum_{i=1}^{q} w_i f_i(\mathbf{x})$$

$$\text{with} \quad \sum_{i=1}^{q} w_i = 1 \tag{5.3}$$

The weighting factors represent the relative importance of the objective functions to the design engineer. It is also common practice to normalize the objective functions so they are expressed in units that have the same numerical importance. For further discussions about multiobjective optimization refer to [54].

5.5   Optimization of Dynamic Performance

The design optimization problems discussed in the previous sections are formulated in terms of time independent quantities, so we can say that the problems involve optimization of static performance of a system. This section goes one step further by considering optimization of dynamic performance, that is, optimization problems where the objective and constraint functions are expressed in terms of time dependent responses from a dynamic analysis of the system. Now the objective and constraint functions may be time dependent. It turns out that by eliminating the time dependence from these functions, optimization problems involving dynamic performance can be solved by the methods described in the previous sections.

In this section we are taking into account that the response vector $\mathbf{s}$ is indeed time dependent, that is, $\mathbf{s} = \mathbf{s}(\mathbf{x}, t)$. The response vector $\mathbf{s}$ may include most of the quantities calculated in a dynamic analysis, that is, time histories of system responses, substructure responses, and natural frequencies of the system. The system responses typically include super node positions, velocities, and accelerations, and forces in super nodes, springs, and dampers. The substructure responses include stress components of individual finite elements. The dynamic performance or behavior of a system is often measured at certain time points or over a selected time interval. In the following, it is assumed that these time points and the time interval are chosen within the fixed time interval of the dynamic analysis.

Recall that the objective function f represents a single measure of design quality or performance of the system. This function is minimized or maximized to improve the design. Further, the objective is a function of the design, directly through the design variables $\mathbf{x}$ or indirectly through the response vector $\mathbf{s}$. In the former case, the objective function can be evaluated once the design variables are assigned numerical values. On the other hand, a dynamic analysis is required to evaluate the response vector and thereby the objective function in the latter case. In general, the objective function may be a function of both $\mathbf{x}$ and $\mathbf{s}$. A typical objective function which is independent of a dynamic analysis is the mass of the system or a substructure. Such an objective function corresponds to the minimum weight problem. In the following we are discussing some objective functions which depend on the dynamic performance of the system.

We may want to evaluate the node position during simulation compared to a prescribed path and the criteria may be the maximum deviation from the prescribed path or the average deviation from the path. If the deviation should be as small as possible, we could formulate this in an objective function or if a maximum deviation is specified this may be specified as an inequality constraint.

Another case could be that we like to have a measure from a single point in space for instance deviation from a point that a robot should pick up something or where it should put something. If minimal deviation from these points is wanted, this should be formulated as an objective function. If the requirement is within a limit from the point, it should be formulated as an inequality constraint.

Along a curve, or at a point, there could be some requirements for instance the lowest natural frequency of the system should be above a specified frequency to avoid excitation of the natural modes of system. This is probably best optimized specifying an inequality constraint.

A minimum time problem could also be the objective for an optimization, for instance how much time will a robot need to move a piece from point A to point B. This could be formulated as an objective function if this time should be as short as possible, but as an inequality constraint if a certain time limitation is specified.

Using KBE for automatic simulation model generation our aim is to pre-establish some generic optimization cases that the designer could use more or less out-of-the-box to solve practical design problems. What is mentioned above could be examples of such "made ready" optimization cases.

5.6 Optimizing Flexible Multibody Systems

In this section we limit the discussion to optimization techniques of practical use when working with KBE for automatic model generation as input to simulation of flexible multibody systems (FMBS). For further discussion of techniques for design optimization refer to references used in this chapter.

5.6.1 Evaluating sensitivities, objectives and constraints

In the KBE application we define objective and constraint functions in terms of time dependent response functions. For simplicity we consider a single response function $q = q(\mathbf{x}) = q(\mathbf{s}, t)$, where $\mathbf{s} = \mathbf{s}(\mathbf{x}, t)$ are responses from a dynamic simulation of the FMBS. Further, we define the design variables within the KBE application. Gradients of q with respect to the m design variables $\mathbf{x}$ are needed to evaluate gradients of the objective and constraint functions. These gradients are sometimes called *design sensitivities* as they can help the designer to improve the design. As shown in Section 5.2, these gradients are used by gradient-based mathematical programming methods to find optimal designs. Below we briefly outline two ways to compute the gradients of q by using m perturbed designs. For the perturbed design $\hat{\mathbf{x}}_i$ we modify the current design $\mathbf{x}$ by adding a small perturbation Δx_i to the design variable x_i, so we have one perturbed design corresponding to each design variable. Thus, the

KBE application may create $m+1$ input files for the current and m perturbed designs.

In the first-order forward finite difference method, or simply the finite difference method, we perform dynamic simulation of these $m + 1$ slightly different designs of the FMBS, and obtain the responses $\mathbf{s}$ for the current design $\mathbf{x}$ and m sets of perturbed responses $\hat{\mathbf{s}}_i = \mathbf{s}(\hat{\mathbf{x}}_i, t)$ for $i = 1, \ldots, m$. In the KBE application, we approximate the gradients of q as

$$\frac{\partial q}{\partial x_i} \approx \frac{q(\hat{\mathbf{s}}_i, t) - q(\mathbf{s}, t)}{\Delta x_i}$$

In the direct differentiation method, we write the gradients of q as

$$\frac{\partial q}{\partial x_i} = \frac{\partial q}{\partial \mathbf{s}} \frac{\partial \mathbf{s}}{\partial x_i}$$

In the KBE application, we have defined the function q in terms of $\mathbf{s}$, so the partials $\partial q/\partial \mathbf{s}$ can be found by analytical expressions (or approximated by a central difference formula within the KBE application). Next, we differentiate the equations of motion to obtain the partials $\partial \mathbf{s}/\partial x_i$ as solutions of a set of so-called sensitivity equations. At each time step of the dynamic simulation, the sensitivity equations are m sets of linear equations, and it turns out that we can use the same coefficient matrix to calculate the responses $\mathbf{s}$ and the m partials $\partial \mathbf{s}/\partial x_i$. The right-hand side of each set of sensitivity equations depends on the corresponding perturbed design $\hat{\mathbf{x}}_i$.

To use the direct differentiation method, we must have access to the source code of the simulation software and extend it so that the dynamic analysis includes the sensitivity equations. This makes the software more complex, and we doubt that the direct differentiation method will be available in commercial software. On the other hand, the finite difference method can be used as a black box where the KBE application supplies the necessary input and receives the requested output. Without parallelization, the direct differentiation method will be more efficient than the finite difference method. The difference method is suitable for parallelization since the $m + 1$ dynamic analyses can be run in parallel; this is more difficult in the direct differentiation method. So if we have enough parallel processors, then the total time when using the difference method will be about the same as the time to perform one dynamic analysis of the FMBS. From the above, the finite difference method seems to be preferred.

Design variables will be specified as selected properties in the KBE application, these may include (i) dimensions, such as positions, angels and lengths; (ii) structural properties such as spring or damper variables; (iii) material properties, and (iv) control variables such as amplification and time constants.

Tools for dynamic simulation of FMBS will usually have a utility for plotting curves for interesting response variables as a function of time or one system variables as a function of another system variable like the x- and y-coordinate of a position for instance to plot a coupler curve. Subtracting,

adding, integration of curves, etc., may also be specified. A very similar specification may also be used to specify optimization objectives and constraints.
The objective function could be to minimize the maximum, the average or the
integrated deviation between one such curve and a reference curve. Similarly,
the deviations could be a measure between a specified point and a reference
point. If a minimum is sought for, an objective function will be specified, if
within some limit is good enough an inequality constraint is specified. If an
exact value is sought for even an equality constraint may be specified. Similarly, deformations, stresses and variations in eigenvalues may be specified as
response curves of interest for optimization in a very similar way as described
above.

5.6.2 Surrogate models and response surfaces

Time simulation of FMBS may, depending on the size of the model, be very
demanding on computational time and from experience time integration in a
simulation code is not well suited for parallel processing because the different
time steps build on each other sequentially and may therefore not be computed
in parallel. The only part that could be sped up to some extent is the solution
of a system of equations that has to be computed one or more times in each
time step. However, the speed-up factor is often quite limited compared to
running everything sequentially. If an optimization process would need many
iterations of the optimization loop and consequently many repeated executions
of the simulation tool, alternative and simplified ways to calculate approximate
simulation results are sought.

Such alternative approaches to generate approximations for the objective
function and constraints are often called surrogate models (SM). Based on a
number of simulations with design variables spanning a feasible area for the
objective function and the constraints, that is judged by the designer to include the optimal solution sought. These preliminary simulations are used to
generate approximate functions for the objective functions and the constraints
to represent the simulation results in the optimization process. These approximate functions are often referred to as *response surfaces* (RS). Optimization
techniques presented in this chapter may also be used in combination with
surrogate models speeding up the optimization process significantly.

When an optimal point for design variable combination is found using the
surrogate functions, you will usually run a full simulation for this approximate
optimum point that may have some deviation from the actual optimal point for
the design. If more optimization is needed, the surrogate model is updated also
using the last full simulation and the optimization process is repeated using
the updated surrogate models. This is repeated until an accepted optimal
solution is found.

If enough parallel computing processes are available for running simulation
concurrently the spanning of the feasible design space with simulations could
be made quite extensively before any optimization algorithms is executed. In

this case the surrogate models could be so accurate that an optimization run based on these models could with high confidentiality represent the optimal solution without rerunning more simulations. Surrogate models based on an extensive number of simulations may also without using optimization be a potential powerful tool for the designer experimenting with variations of design variables in the search for an improved design. In the next section we will discuss in more detail how surrogate models could be used for optimization of FMBS in a KBE environment.

5.7 Tool for Optimizing Flexible Multibody Systems

This section is to a large extent based on the work Arnt Underhaug Lima [25] did for his master's thesis.

Based on quite extensive evaluation and testing the selected framework for optimization of FMBS is OpenMDAO [38, 21]. This framework has model decomposition capabilities, interfaces to a variety of optimization algorithms, includes multiple surrogate model algorithms and incorporates multiple designs of experiment strategies. The benefits of using this framework include a modular approach both for the optimization strategy and modeling of the problem. For example, changing the optimization algorithm is a simple one-line change. The model decomposition also makes it possible to define the flow of parameter across model components, without having to define the execution order. For this application OpenMDAO is integrated with the optimization algorithms SLSQP, SNOPT [18] and IPOPT [65]. The former two of these are sequential quadratic programming (SQP) algorithms; the latter is an interior point (IP) method; for details of the IP method refer to [34]. SNOPT is a commercially available algorithm, whereas the other two are released under open-source licenses.

The main reasons for selecting these algorithms are as follows. IPOPT and SNOPT have been claimed to be the state-of-the-art algorithms for optimizing nonlinear problems [34]. The same algorithms have been shown to have superior performance for engineering problems, especially over genetic algorithms [47, 66]; for this reason any use of genetic algorithms has not been considered. Finally they provide, as mentioned, samples of both SQP and IP methods, which means the two algorithms are reasonably different. The optimization loop is presented in Figure 5.2.

The program systems shown for the optimization loop in this figure are commented briefly in the following. The optimization is controlled by the OpenMDAO Design Optimizer pyOpt, an open source tool implemented in the programming language Python. pyOpt supplies the design parameters as input to the KBE application implemented in AML and is the basis for automatic generating of the geometry for the system to be designed. The next step

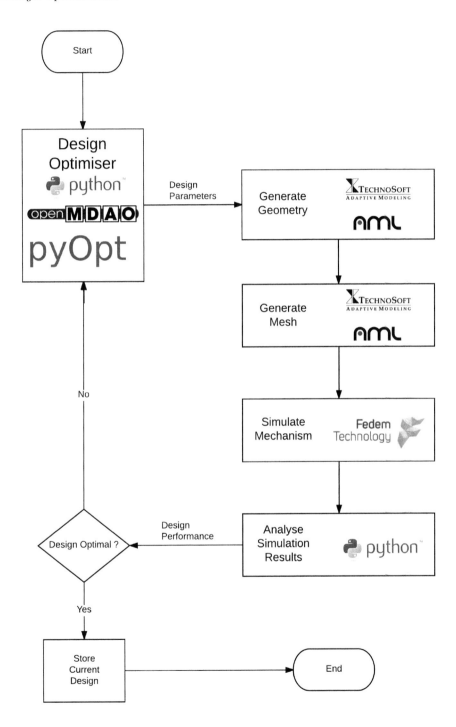

FIGURE 5.2: Software utilized in the optimization loop [25]

in the KBE application is automatic generation of FE meshes for each part of the mechanism assembly. An individual file with FE data is generated for each link in the NASTRAN Bulk Data Format (BDF). These files include rigid connection points for external entities like mechanism joints. The KBE application also generates the system input for the mechanism system referring to the BDF files for the mechanism links and adding other system components like springs, dampers, loading, etc., to define the simulation model completely. With a complete simulation input dataset, the simulation tool FEDEM [3] is started. The response results from the FEDEM simulation are analyzed in the Analyze Simulation Results module implemented for this application using the programming language Python. The analyses of the simulation results are used to calculate values for the objective function and the constraints. If an optimum point is not reached, a new optimization iteration is initiated and the optimization algorithms in pyOpt calculate new values for the design parameters and the processing described above is repeated. The iteration is terminated when an optimal design is found based on the formulated optimization criteria defined. The optimization loop described here is what we will call *direct optimization*.

If surrogate models are used, the optimization loop is somewhat different than in Figure 5.2. The first step will be the same as for direct optimization in order to define design parameters with move limits, defining the objective function and defining the constraints. For direct optimization the optimization loop will be the next step, but running with surrogate models the next step will be to generate sets of values for design parameters varying each design parameter between its move limits in an orderly fashion to cover the whole design space. The next step will, for each set of design parameters, be to generate geometry and FE mesh in the KBE application, to run simulations and to calculate values for the objective function and for the constraints. This is repeated for each set of design parameters defined. With all simulations completed, the surrogate models are generated for the objective function and the constraints. With the surrogate models, also called response surfaces (RS), available the next step will be the optimization. However, running optimization based on surrogate models the KBE application and the simulation tool will not be involved, that is the top three blocks to the right in Figure 5.2. The analysis will in this case only be to evaluate the objective function and the constraints, including any sensitivities needed from the response surfaces, and from this the optimization algorithm pyOpt will generate new sets of design parameters for new evaluation of objective function, constraints and sensitivities. This sequence repeats itself until an optimal design is found based on the defined criteria. If the design space is covered well for all design variables prior to generating the surrogate models, the resulting optimal design from this process will be very close to the optimal design for the physical system. However, you will probably run the KBE application and the simulation tool over again for the found optimal solution for verification purposes.

6

Environment for Design Automation

The KBE examples and applications in this book use the Adaptive Modeling Language (AML) framework as the selected tool supported by TechnoSoft Inc. [60, 22] in Ohio, USA. The KBE applications generate input to the FE-DEM simulation tool supported by the FEDEM Technology Company [3] in Trondheim, Norway.

6.1 KBE Development Framework

Originally AML was built on top of the Lisp language, but was later implemented in the language C as a standalone compiler. As described in chapter 2 for KBE languages in general, the AML language is based on the object oriented (OO) paradigm. For more than 25 years more than 100 industrial KBE applications have been developed in AML. With the historical basis in the Lisp language, operations on lists is both a key and powerful feature in AML. The most important feature of AML compared to traditional OO languages is the close integration with geometric libraries and geometric operations. A very wide span of KBE applications has been implemented in AML with the aim of automating engineering and design work. Applications span automotive systems, aviation systems, marine and offshore systems, agriculture systems, etc. Examples of operations that have been automated are generation of structure geometries, routing of cables, piping, wires and ducts, generation of storage tanks, layout generation and more. What is common for all these applications is that the implementation is based on detailed studies of work processes and communication with engineering experts doing this work manually.

Design is always closely integrated with analyses of different types spanning from manual calculations to full-scale finite element analysis of whole systems. Manual analysis is usually implemented directly in the KBE application, but integration with external analysis systems is an important part. A typical kind of external analysis is finite element (FE) analysis. The steps for doing FE analysis could be to transfer the geometry generated in AML, as for instance a STEP file, into a FE preprocessor for FE mesh generation and subsequent FE analysis in a standard FE package, for instance NASTRAN, ANSYS, ABAQUS, and others. However, an even more efficient way of doing

FE analysis in an AML application is to do the FE meshing internally in AML where automatic mesh generation for 2D and 3D geometries is available. With this functionality complete FE analysis input could be generated automatically in the AML application for instance as standard NASTRAN bulk data format (BDF) files. The FE analysis could then be started remotely from the AML application, and the results of the FE analysis are imported back into the application for inspection.

There could be cases where a human inspection of analysis results is not the most efficient way; for instance where it could be very difficult for even an experienced designer to see what would be the best changes to make in order to achieve the requested improvements in the design. In these cases automatic optimization should be considered. Techniques for automatic optimization have been available for many decades, but one should not overestimate how easy it will be to implement this as a viable solution for automation in design. This optimization option is available in AML in a module called AMOpt, and we think it should be utilized in some design situations. In this book we will discuss ways to use automatic optimization in connection with FE analysis. In most cases automatic optimization should be used in combination with manual optimization input. We don't believe that eliminating the human designer from the design loop is a viable solution. As discussed more in the optimization chapter (Chapter 5), the main challenge of implementing optimization techniques in the design loop is for the designer to specify the objective function that controls the design loop. Besides, will such an objective function be static as the design process progresses, or should the objective function change for the different stages of the design? The optimization loop will include a parameterized model of a design in AML where some of the parameters are selected as design variables which could be changed manually or automatically through design optimization. For each design iteration, FE input will be generated and FE simulation carried out. The results from the simulation will be the input to the objective function, or the designer for the manual optimization alternative. The optimization algorithm, or the designer, will select a new set of values for design variables in the AML model and the process is repeated. When some optimization criteria are fulfilled, the process is terminated.

Typical for a KBE application is that in the early concept phase for a design, the KBE application also generates the details of the design. This means that all alternative concepts could be studied in detail before any decision about selection of concept has to be made because the detailed alternatives can be generated in minutes or seconds. This means that the design process will be very different using a KBE application since early and late design phases are combined. The designer doesn't have to make design decisions based on preliminary and rough concept models, but could inspect the consequence of design decisions at any level of detail.

6.2 Boolean Geometry

AML is integrated with the geometric modeling kernel PARASOLID [55]. A number of CAD systems are also based on PARASOLID. In a CAD system, the interaction with the PARASOLID kernel is achieved through an interactive graphical interface based on the desired picture of the geometry in the designer's head, i.e., the knowledge used to define the geometry is not codified any place besides in the head of the designer and in the graphical model displayed by the CAD system. In a KBE language like AML, the geometry is codified and parameterized in a programed algorithm. The parameters that generate the geometry are usually not set directly, but generated by traversing some engineering rules. This is what's called generative design in KBE. The KBE system does not store the geometry itself, but a recipe for how to generate the geometry. In a CAD system you could have a parameterized geometry, but you will not have stored the rules for how to set the geometry parameters. Using the programming concept of the KBE system, based on some engineering rule variation, you could switch to a different recipe for geometry generation, and end up with a quite different geometry as a result; both in topology and dimensions. The Boolean geometry operations implemented in AML are described briefly in the following sections.

Boolean operations are based on geometric primitives as a box, a cylinder, a cone, etc. More complex primitives are generated from a more or less complex surface which is extruded into a volume (or a line extruded into a surface, or a point extruded into a line). Having two or more of these types of geometric entities positioned in space, a number of Boolean operations may be carried out.

A union operation combines the geometric entities referred to in a list into one new continuous geometric entity without overlaps. A difference operation takes the first geometric entity in a list and subtracts the geometries for all the following geometric entities from the first geometric entity. This will be the resulting geometric entity. An intersection operation makes a new geometric entity from the volume where all geometric entities in a list overlap. Resulting geometric entities from any of these Boolean operations may be involved as geometric entities in new Boolean operations.

The AML source code for generating a binary link is shown in Tables 6.1 and 6.2 and the resulting geometry is shown in Figure 6.1. In Table 6.1, the binary-link-union-class is used to glue a cylinder between two points to two 90 degrees rotated cylinders in each end as a first step in generating the binary link. The first subobject in the binary-link-difference-class (see Table 6.2) is referring to the binary-link-union-class for the first object in the object-list, and from this object cylinders are subtracted in both ends to make the holes in the joints; see Figure 6.1.

In addition to geometries generated from Boolean operations, general

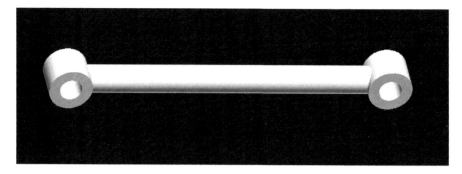

FIGURE 6.1: Binary link geometry

sculptured surfaces may be generated in AML that are used to model smooth surfaces of cars, airplanes, ships, submarines, etc. These types of surfaces may also be involved in Boolean operations to generate very complex and compounded geometries.

6.3 Automatic Meshing

The aim of this book is to present a method to automate as much as possible of the process of dynamic simulations of mechanisms. To do a high fidelity simulation of a mechanism, detailed geometry of each part (link) of the mechanism is needed in addition to the connectivity between the links as joints, springs, dampers, control elements, loads, etc. For each link, a separate input file to the simulation tool is needed with the finite element mesh generated for the actual geometry of that link. Typically these files for the different links will be generated in the standard NASTRAN bulk data format (BDF). AML has functionality for automatically generating tetrahedral finite elements for any solid geometry modeled in AML. For each link a tetrahedral FE mesh is generated, and a BDF file is generated for that link ready as input to the simulation tool. If a link is better approximated by a surface, triangular elements may be generated automatically for that surface and also produce a BDF file for input to simulation.

Any point in a FE mesh could potentially be a connection for a system element like a joint, spring, damper, control element, load, etc. However, external nodes in a link referred to as super nodes are positioned more or less by chance by the meshing algorithm and not positioned exactly where we like to connect to system elements. Besides, connecting a super element to a single node may produce stress singularities around that point. The best way to solve this is by using multipoint constraints (MPC). One such MPC is named RBE2 in

TABLE 6.1
Binary link union class in AML.

```
(define-class binary-link-union-class
  :inherit-from (union-object)
  :properties(
;;;              length-link  1.0
;;;              node-0-coords '(0 0 0)
;;;              node-1-coords (list (append ^length-link) 0 0)
             object-list (list ^joint-cylinder-0 ^joint-cylinder-1
                               ^link-cylinder)
             )
  :subobjects(
             (joint-cylinder-0 :class 'cylinder-object
                 diameter ^^joint-outer-radius-0
                 height ^^joint-0-height
                 orientation (list (translate ^node-0-coords))
                 )
             (joint-cylinder-1 :class 'cylinder-object
                 diameter ^^joint-outer-radius-1
                 height ^^joint-1-height
                 orientation (list (translate ^node-1-coords))
                 )
             (link-cylinder :class 'cylinder-object
                 diameter 0.03
;;                  height  ^length-link
;;                  orientation (list (rotate 90.0 :y-axis)
             (translate (list (append (/ ^length-link 2)) 0 0)))

             Height (vector-length
                     (subtract-vectors ^^node-1-coords ^^node-0-coords)
             orientation (list (align (center-of-face (the cylinder-object) 0)
                     (normal-to-face (the cylinder-object) 0)
                     nil
                     ^^node-0-coords
                     (subtract-vectors ^^node-1-coords ^^node-0-coords)
                     nil
                     :move? t :align? t :orient? nil
                     )
                 )
             )
      )
  )
```

TABLE 6.2

Binary link difference class in AML.

```
(define-class binary-link-difference-class
  :inherit-from (difference-object)
  :properties(
              node-0-coords (default '(0 0 0)) ;;(nth 0 ^^coords-list)
              node-1-coords (default '(1 0 0)) ;;(nth 1 ^^coords-list)
              joint-outer-radius-0  0.05 ;;(default 0.05)
              joint-inner-radius-0  0.0255 ;;(default 0.025)
              joint-outer-radius-1  0.05 ;;(default 0.05)
              joint-inner-radius-1  0.0255 ;;(default 0.025)
              joint-0-height 0.04 ;;(default 0.04)
              joint-1-height 0.04 ;;(default 0.04)
              joint-0-hole-height 0.1 ;;(default 0.1)
              joint-1-hole-height 0.1 ;;(default 0.1)
              object-list (list ^binary-link-union ^joint-cylinder-hole-0
                                ^joint-cylinder-hole-1)
               )
  :subobjects(
              (binary-link-union  :class 'binary-link-union-class
               )
              (joint-cylinder-hole-0 :class 'cylinder-object
                 diameter ^^joint-inner-radius-0
                 height ^^joint-0-hole-height
                 orientation (list (translate ^node-0-coords))
                 )
              (joint-cylinder-hole-1 :class 'cylinder-object
                 diameter ^^joint-inner-radius-1
                 height ^^joint-1-hole-height
                 orientation (list (translate ^node-1-coords))
                 )
               )
  )
```

the NASTRAN BDF file. A RBE2 is a rigid connection between a number of FE nodes and the node position where the system connection is planned to be. Ideally all the points on a link that are planned to be connection points at system level should be generated as RBE2 points. The RBE2 points will be super nodes on the actual link. The theoretical formulation for MPCs is presented in Section 3.7.

7

Interaction with the User

A KBE application for automated generation of input data for mechanisms simulation must have a good interface for the user to edit input parameters and execute simulations. The parameterized input is implemented as files in a mechanism library which the user can select from and run simulations on. New mechanisms may be entered into this library by editing and entering alternative and/or additional files into the library. The menu for top level mechanism selection from the library, visualization for geometry and FE meshes, editing of parameters and exporting of input data to the simulation code is shown in Figure 7.1.

7.1 Entering and Editing the Mechanism Library Format

The content of this section is to a large extent based on the work of Thor Christian Coward [10] in his master's thesis.

The mechanism library format presented in Section 4.2 is the basis for entering new mechanism topologies into the mechanism library implemented for the Rapid Mechanism Modeling System (RaMMS) presented here. A text editor like Notepad++ is well suited for entering and editing files or lists with data like *node positions, constraints, link shapes, springs and dampers, loads*, and other optional definition files. The mechanism library for the KBE application RaMMS will include topology definitions for a number of typical mechanisms like linkages, suspension systems, cranes and robots, windmills and others. Each mechanism type will have one or more feasible example *node position* file that could be selected for simulation input generation, and subsequent execution of dynamic simulation. The node positions may easily be edited interactively and stored as a new version in the library, and new simulation input generation may be executed.

Below the *library* folder for the application, a number of different mechanism folders are available with distinct names describing the mechanism at hand. Below this mechanism type folder level there is another folder level for versions of the actual mechanism, and at this level the actual mechanism definition files are stored; refer to *node positions, constraints, link shapes, springs*

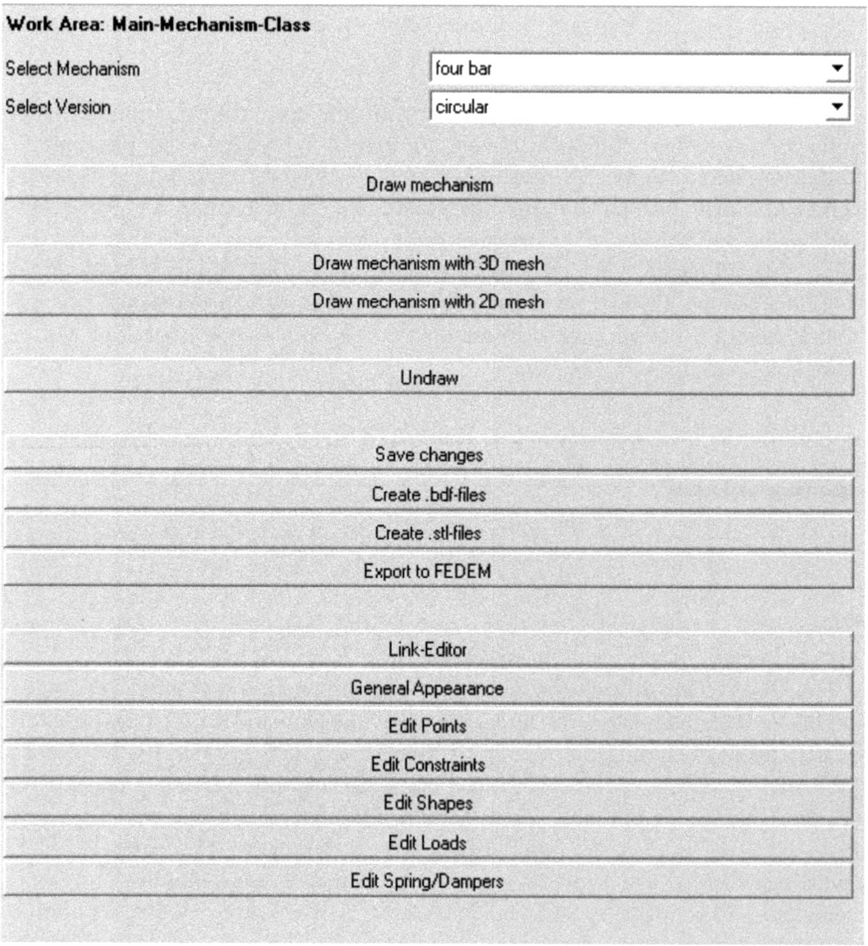

FIGURE 7.1: Overall user menu including selection from and export to library [10]

and dampers, *loads*, and other optional definition files. The different versions will typically have different numerical values for global coordinates and/or different cross section types for members in links and/or different numerical values for the member cross sections.

Editing the library files as described above is absolutely a workable solution, but entering and editing these data through a graphic user interface would be much more efficient and attractive for the user. There are two levels of input data to the KBE application presented here: The primary data that has to be entered to RaMMS is the data that has to do with positioning in space of, for instance, joints, link topology and link geometry and dimensions. This data is used directly to generate the detailed geometry for links and joints including blending of sharp edges between link member and joint geometries. This detailed geometry is the basis for automatic generation of surface or volume FE meshes for the different mechanism links. For each link the mesh generated is the basis for generating a file in Nastran Bulk Data Format (BDF) including rigid connections (RBE2) for joint; for details about RBE2 connections refer to Section 3.7. The main work gain when doing dynamic simulations in FEDEM is achieved from this primary input described above. In cases where design iterations are controlled manually, an option could be to enter only this primary KBE input to the KBE application, and then enter and edit other simulation input directly through the FEDEM user interface.

If the design iteration will need several iterations through the RaMMS KBE application, it could be advantageous to enter more simulation input through the KBE application in order to avoid redoing extra simulation input in the FEDEM interface for each design iteration. When doing automatic design optimization it will be necessary to enter all simulation input to the KBE application because no manual simulation input modeling will be possible in an automated simulation loop. The challenge in this case is that you would need to redo the FEDEM simulation input functionality fully for the KBE application. A somewhat reduced version of the full simulation of input modeling for the KBE application is chosen here in addition to the primary input data like joint positions, mechanism topology and joint shapes. Springs, dampers and loading are also included in the KBE application input. Input like for instance control system modeling is not included in the current version of the pilot KBE application.

7.1.1 Menus for entering/editing the KBE library format

The positioning of the mechanism keypoints (Nodes) is entered in the Keypoint modeling window 7.2 with a pop-up widget containing a table for entering coordinates. For easy visualization when defining keypoints, a symbol and keypoint number will appear at the correct position in the modeling window, when the user is finished defining the keypoint, and jumps to the next line in the table.

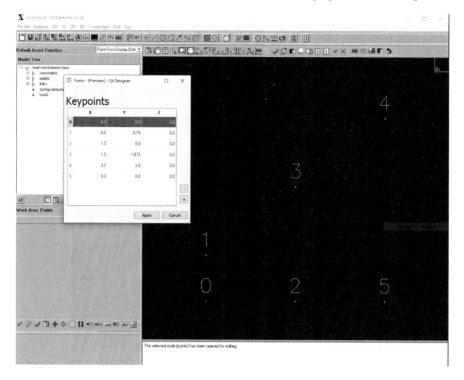

FIGURE 7.2: Keypoint modeling menu [10]

Joint and link topology is entered through the joint modeling window 7.3 with a widget containing a table for entering keypoint number, joint type, links involved and joint direction. When a line is completed in the constraints widget, keypoint numbers appear in the modeling window. When the user has defined one constraint, and switches line or presses return, a symbol representing the joint will be shown in the modeler. Joints on the same link will be connected by a line. For the joint you are editing the joint symbol turns red.

Springs and dampers are other key elements in many mechanisms, and are entered by the Spring and damper menu; see mock-up in Figure 7.4. You could define axial springs and dampers between points on different links, or you could define springs or dampers on joint degrees of freedom.

Another important mechanism input for dynamic simulation is the loading of the system and this is entered through the loading menu; see mock-up in Figure 7.5.

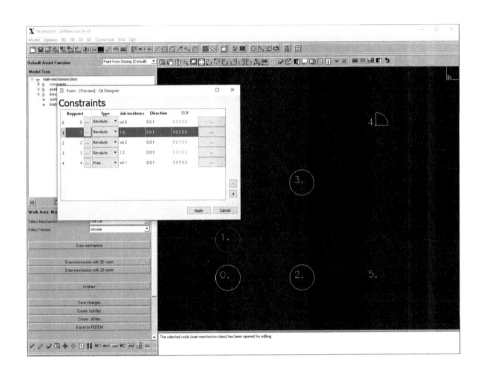

FIGURE 7.3: Joint modeling menu [10]

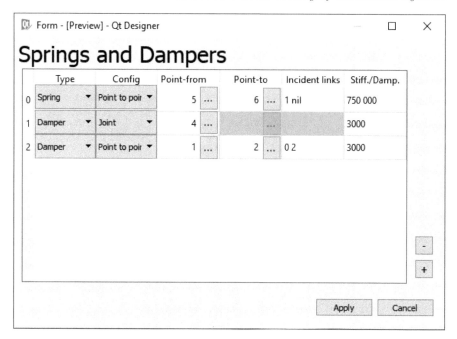

FIGURE 7.4: Spring and damper modeling menu [10]

7.1.2   Link editing module

Link shapes could be edited directly in the link shape library file; see Table 4.3. However, for complex link geometries in 3D space where curved members are required, a more advanced modeling tool is needed. To ease the link modeling task, a graphic link editor is developed; see Figure 7.6. The library link shape format can be used to model quite complex link geometries, but doing this manually through a text editor can be quite demanding and often requires some iterations including graphic displays for each iteration.

This widget has its own separate modeling window displaying only the selected link. If the link is ternary or higher degree, the user modifies the members individually in the modeling window. The selected member will be a red line, while the other members of the link will be white lines. Member geometry can be removed from the link by ticking the "hide" box. Spline curve weight points for a member can be added, deleted, modified and moved interactively to form desired member shapes when extruded along the spline curve. Member cross sections can be set also with varying cross section from member start to end. Also the member mesh size can be specified. The values for cross section and mesh sizes will be displayed; these are values calculated by the system, which the user can override. Blending of sharp edges between member and joint geometries can be switched on and off, and the blend radius

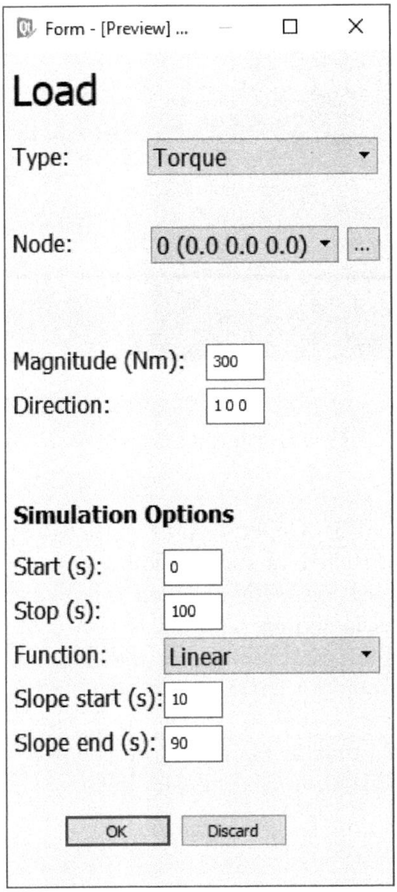

FIGURE 7.5: Load modeling menu [10]

)

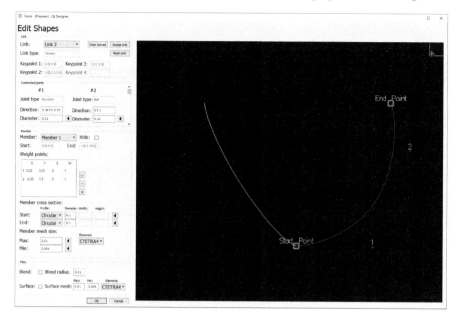

FIGURE 7.6: Link editing viewer [10]

can be set in the link editor. Generation of surfaces between closed loops of members may also be switched on and off. The link can also be meshed in the link shape editor for mesh inspection. When the geometry, potentially with blending of sharp edges including meshing is tested in the link editor, it is ready for simulation input generation, and the edited link is completed and written back to the library format and the next link is entered for editing. The Link editor has been tested for editing a Double Wishbone suspension system; see the meshed result in Figure 7.7. In this way the link editor works as a customized CAD system for mechanism links.

7.2   Other Simulation Input

Simulation inputs that are not modeled for this first version of the RaMMS KBE pilot, but should be implemented in a later version, are discussed in this section. Simulation parameters like time integration end-time, time steps and convergence tolerances, other than default values, are typical parameters that should be considered. Also the simulation tool may specify eigenvalue and eigenvector calculations for certain time steps during simulation. Vibration modes at certain positions and with different loadings are also important input

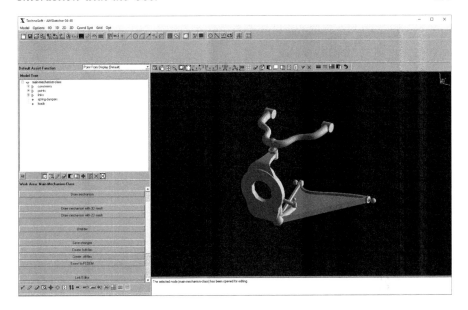

FIGURE 7.7: Double wishbone suspension edited and meshed [10]

to the designer when making design decisions. A typical effect that could be important is how for instance the eigenmodes for a suspension system change when it is loaded and the momentary stiffness of nonlinear springs changes, for instance in rubber bushings in bearings. This is information that could be important during a manual or automatic design optimization iteration. Modeling of functions is another important input to simulation. They are used for specifying nonlinear spring stiffnesses and damping coefficients, change in stress free lengths or angles of springs, variation of input forces and torques in the model, and last but not least reference functions for comparators in control systems; see next paragraph.

Modern machines are very seldom a pure mechanical thing, but rather integration of mechanical parts with sensors and actuators interacting with a control engineering system. When optimizing the system, the control parameters should be updated to achieve an overall optimum system and avoid the over-the-wall situation where the mechanical engineers develop and optimize the mechanical part of the system and at a later stage the control engineers develop the control system that should complete the combined mechanical, electrical and hydraulic system and with little feedback between the different experts. Control engineering modeling means modeling of entities like:

- Control input and output that are connections to sensors and actuators in the system

- Amplifications and time constants

- Comparators, adders and multipliers

- Integrators and derivators

- Sample and hold and delays

- Logical switchers and limiters

- Dead zone and hysteresis

- PID, PI and PD controllers

- Real pole and complex conjugate poles

- 1st and 2nd order transfer functions

To develop such a control editor specifically for RaMMS seems to be a little bit of overkill. Integrating the control editor from FEDEM into RaMMS would be a better approach, especially as it is a quite independent module of the FEDEM user interface.

7.3   Design Parameters and Optimization

When we close the loop from primary input data for positions and topology in a mechanism, and include automatic meshing and generation of simulation input, we are quite close to completing an automatic optimization loop. This is for mechanisms where optimization algorithms update simulation parameters and iterate towards an optimum design based on some objective function, and satisfy defined optimization constraints. To do this without any custom-made programming input should be feasible, but is a quite demanding task with regard to implementation effort, especially to implement an intuitive user interface. To develop a generic optimization user interface for all kinds of simulation input and mechanism topology seems to be a quite extensive task, but to develop custom-made optimization input for automatic optimization of classes of mechanisms could be a more reasonable approach, for instance to:

- optimize a four-bar mechanism for the coupler point to follow some prescribed curve should be quite easy to implement without any programming effort

- optimize a Stephenson 3 mechanism for some specified handling operation between two specified positions with some velocity and acceleration requirements should be quite easy to implement without a programming effort

- optimize for instance a McPherson front suspension for some key suspension angles (caster, camber, toe radius, etc.) using the position of the top structure as design variable could be a potential task for automatic optimization without programming effort from the user.

The challenge for automated optimization, as it has always been, is for the designer to formulate the optimization objective function and constraints mathematically, and in addition to program this as input to the optimization algorithms.

Developing optimization for mechanism design we have been testing what is called direct optimization that is running the geometry generation in RaMMS, including meshing and generation of simulation input, followed by a simulation and then evaluating the simulation results for the objective functions and the constraints, as well as finite difference calculations for sensitivities with respect to selected design parameters in order to generate an updated set of design parameters to RaMMS for the next optimization step.

An alternative approach is to use surrogate models for the optimization. In this approach we select design parameters with upper and lower move limits and generate a set of simulation inputs by stepping the design parameters between the move limits in an orderly way. We then run a quite large number of independent simulations that cover the design space quite well. This will require quite large computational resources, but as these simulation runs are independent, they may be run in parallel and the clock time does not need to be much more than for one simulation run. The advantage of this is that you have effective tools for generating response surfaces for the objective function and for the constraints. The input to the optimization algorithms is generated very effectively, and searching for an optimum is done without any more execution of RaMMS and the simulation code. The found optimal solution could be rerun through RaMMS and the simulation code, but with quite extensive simulation runs over the design space to generate the response surface, the solution selected by the optimizer should be very close to the real optimum solution for the design.

A third approach to design optimization could be to run a quite large number of simulations scattered over the design space and instead of using automatic optimization manual optimization could be used to search in the design for an improved solution using sensitivities with respect to design parameters and the value of the objective function and the constraints. Also in this case design variables, objective function and constraints need to be selected to generate the response surfaces, but the formulations do not need to be so complete as for automatic optimization because the designer's judgment could more easily be added to compensate for any lack of consistency in the mathematical formulations. Having run a comprehensive set of simulations to begin with, formulation of objective function and constraints could be edited from experience with the first version and new response surfaces could be generated for continued manual search for optimum or the first manual search could be used for tuning of the formulations to continue with automatic search for optimum using optimization algorithms.

The forth alternative approach could be to run the design optimization manually using the designer's intuition to change input parameters to RaMMS, generate simulation input and run simulations.

All the three first alternatives described above require that the designer select design parameters and formulate objective function and constraint functions. Selecting design parameters and moving limits are tasks where the designers insight into the problem is needed. Formulating objective function and constraints mathematically requires even more from the designer's insight. This is the main reason that automatic optimization to a very little extent has been used in design of complex mechanical systems. With the help of the RaMMS system, it will be possible to explore optimization alternatives in mechanism design as described above. Closing the design loop from primary design parameters into RaMMS to full dynamic simulation has large advantages in all situations described above because it dramatically reduces design work with simulation in a design iteration; see also Section 5.7.

A case for automatic optimization of mechanisms is presented in Section 8.5.

7.4 Extended KBE Programming

For the user of the RaMMS application, there will in many cases be requirements, or at least intentions, from the user or designer to do for instance geometric modifications that are difficult or impossible to do with the tools available in RaMMS. One option could be to develop the customized CAD modeler in the Link editor further to include more geometric modeling entities based on the PARASOLID [55] kernel included in AML. Another alternative could be to open up the AML code in the Link editor to programmatically modify or rebuild the link geometry by including customized AML code. The third alternative would be to implement an API to the RaMMS application to modify or replace models inside RaMMS. The most generic alternative could be to develop STEP translators in RaMMS to be able to export and import geometric models for links to and from a general purpose CAD system. The AML framework has utilities available to support potential implementation of STEP translation interfaces.

8

Automated Design Cases

This chapter is based on a pilot KBE application for automatic generation of mechanism simulation input called Rapid Mechanism Modelling System (RaMMS). The last section in this chapter shows how automatic optimization may be used to improve the geometry and the parameters for a mechanism to fulfill functional requirements (e.g., synthesis) and structural integrity requirements.

8.1 Linkages

In everyday language linkage is almost a synonym for the term mechanism. Unfortunately (or fortunately, depending on point of view) the term mechanism is today used in many situations with no reference to constrained mechanical systems in motion, as we always refer to in this book when using the term mechanism. There could be many reasons why this is so, but one could be that when you use the term mechanism you refer to something that is well defined, but is difficult to explain in detail using few words. Our use of the term mechanism is therefore constrained mechanical system in motion.

Linkages have often been used for handling parts between different machines in a production factory. The advantage of using linkages for handling purposes is the precision and repeatability you can achieve, and the speed with which the handling can be done. The disadvantage is the complex task of designing a mechanism that can do the desired handling, i.e., the synthesis problem, and the lack of flexibility when changes the handling process are required. Using a robot for the handling, which is often the case today, you get the flexibility to change the handling just by changing a program, but the speed and precision of the handling would usually be poorer than for a custom designed linkage. Speeding up and automating the design process for linkages could potentially change the balance when you would choose a linkage instead of a robot for a particular handling task. The more mature and stable the handling process is, and if the goal is to speed up the process, the advantage of using a linkage would be clearer. The approach could be to use robots for handling in a new production facility when the process is still developing, and

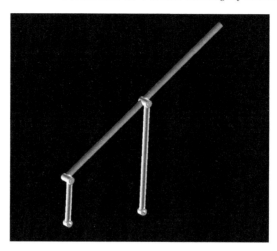

FIGURE 8.1: The four-bar straight line generator

then change from robot to linkage when the handling is stable, but needs to be speeded up to increase production.

8.1.1   Four-bar mechanisms

Maybe the most used linkage in industrial applications is the four-bar mechanism, here represented by the Four Bar straight line generator; see Figure 8.1. The input lists in Tables 4.1, 4.2 and 4.3 of Chapter 4 are using the data for this mechanism for demonstration of the list formats. Figure 8.2 shows the mechanism with all three links meshed with solid elements.

To demonstrate how springs and dampers are modeled, one spring and one damper are defined; see Table 8.1. From this table we can see that an axial spring is connected between node 5 on the coupler link (1) and node 6 on the ground link (nil). The spring stiffness is 100 $\frac{N}{m}$ (a fairly low stiffness). As also may be seen from the table, an axial damper is connected between node 1 on the crank link (0) and node 2 on the rocker link (2). The damper coefficient is set to 5 Newton second per meter. To drive this mechanism, a torque is applied in the crank joint simulating input from an electric motor; see Table 8.2. The torque is applied in the crank bearing (Node 0) in negative z-direction and with a magnitude of 75 Nm. Figure 8.3 shows the four-bar mechanism exported to FEDEM where also the damper and the spring modeled in Table 8.1 are displayed. Animations after running simulation in FEDEM for 10 seconds with the input data described here, produce two and a quarter clockwise rotations of the crank. Stress calculations waere specified from 9 to 10 second simulation time. Stress contour plots after 9 seconds are shown in Figure 8.4 and the stress

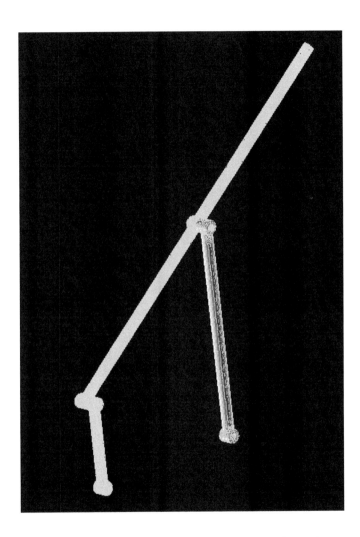

FIGURE 8.2: The four-bar mechanism with mesh

TABLE 8.1
Four-bar spring/damper input list [24].

Type	Node-start	Node-end	Links	Stiffness/Damping
"Spring"	5	6	(1 nil)	100.0
"Damper"	1	2	(0 2)	5.0

TABLE 8.2
Four-bar loads input list [24].

Type	Node	Direction	Magnitude	Link
Torque	0	(0.0 0.0 -1.0)	75.0	0

contour plots after 10 seconds iare shown in Figure 8.5. These results may of course be shown as animations with or without stress contour plots included.

Editing the top rocker position 0.2 m to the right in the RaMMS KBE application new simulations may be carried out by automatically generating new link geometries, new link meshes and new simulation system input. The modified start position for the mechanism is shown in Figure 8.6, and the end position for the simulation is shown in Figure 8.7. Here the mechanism has rotated a little more than two and a quarter clockwise rotations.

8.1.2 Six-link mechanisms

Two types of six-link mechanisms are very familiar, the Stephenson six-link (see Figure 8.8) and the Watt six-link (see Figure 8.9).

The Stephenson mechanisms consist of 2 ternary links connected by two binary links and a dyad (two binary links in series). For the Stephenson-1 mechanism one of the connecting binary links is ground, for the Stepenson-2 mechanism one of the binary links in the dyad is ground, and for Stephenson 3 one of the ternary links is ground; see Figure 8.8. The topology is the same for all Stephenson mechanisms, the only difference is which link is the ground link.

The Watt 1 and 2 mechanisms are shown in Figure 8.9. The main difference between the Stephenson mechanisms and the Watt mechanisms is that the ternary links are directly connected in a joint, and that the ternary links are also connected with two dyads. For the Watt-1 mechanism one of the binary links is ground and for the Watt-2 mechanism one of the ternary links is ground.

Designing with these six-link mechanisms is mainly to move around the joint positions typically to achieve a certain motion. This positioning of joints could be done manually, or the joint positions could be generated in a mechanism synthesis program. In the MECSYN synthesis program [33] you can specify in detail the motion (for instance positions, velocities and/or accelerations) you need and the program will calculate the joint positions for you. One

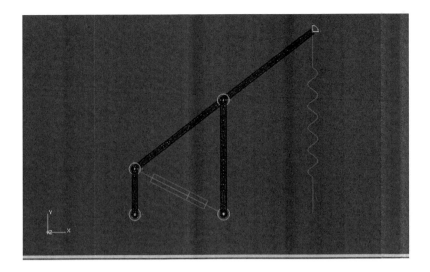

FIGURE 8.3: The four-bar mechanism in FEDEM

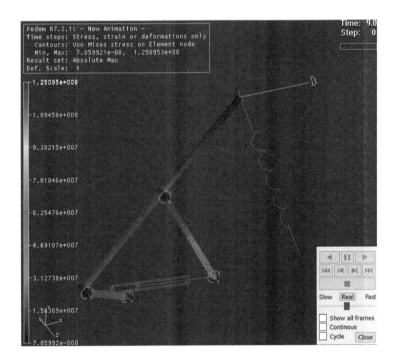

FIGURE 8.4: Stress contour plots after 9 second's simulation

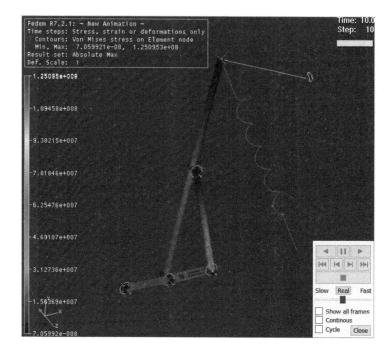

FIGURE 8.5: Stress contour plots after 10 second's simulation

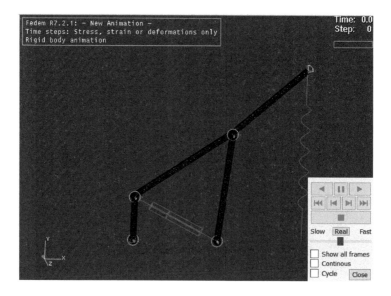

FIGURE 8.6: Position of top rocker joint edited 0.2 meters to the right

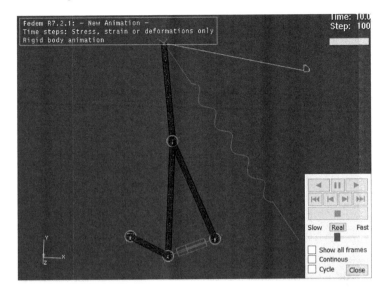

FIGURE 8.7: End position for modified simulation input

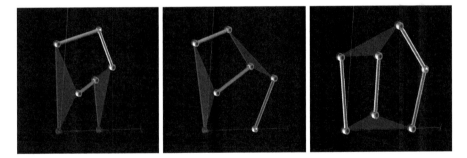

FIGURE 8.8: Stephenson 1, 2 and 3 six-link mechanisms

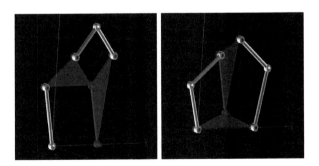

FIGURE 8.9: Watt 1 and 2 six-link mechanisms

TABLE 8.3

Synthesis specifications in metric units [33].

l	j	k	$x_j^{(k)}$		$y_j^{(k)}$		$\phi_j^{(k)}$	
0	0	0	0.79248	$[m]$	0.27432	$[m]$	$\pi/2$	$[rad]$
1	0	1	0.86210	$[m/s]$	0.86210	$[m/s]$	π	$[rad/s]$
2	0	2	2.70839	$[m/s^2]$	-2.70839	$[m/s^2]$	0.0	$[rad/s^2]$
3	1	0	0.4572	$[m]$	-0.3048	$[m]$	0.0	$[rad]$
4	1	1	0.3048	$[m/s]$	0.0	$[m/s]$	0.0	$[rad/s]$

common problematic material handling task is transferring an object from a continuous moving conveyor belt to a continuously rotating turret. Example mechanisms for this handling problem are synthesized in [33, 31, 32]; see figure 8.10.

In Table 8.3 the synthesis specifications are presented in metric units. For this dynamic synthesis approach position, velocity and acceleration are given for the drop off position (line 0 to 2 in the table), and position and velocity for the pick up position (line 3 and 4 in the table). Table 8.4 shows the input data to the RaMMS application for nodes and constraints input, respectively. For link shapes, the same default input is used for all links as in Table 4.4 for the four-bar mechanism example except the dimensions here are (0.04 0.04). The automatic generated and solidly meshed six-link mechanism by the RaMMS application is shown in Figure 8.11. Two positions from the simulation of this mechanism in the FEDEM simulation tool is shown in Figure 8.12. The coupler point position and velocity variation are shown in the graphs presented in Figure 8.13.

8.2   Suspension Systems

A second very large class of mechanisms is suspension systems. Suspension systems are made to protect some vehicle or content of a vehicle from damage when the vehicle is moving on a rough surface and/or giving people occupying the vehicle a pleasant ride. There is a very large variation of vehicles that need suspension systems as for instance passenger cars, trucks, buses, trains, construction machines, tractors, landing wheels for airplanes and for each of these classes of vehicles there is a large variation of suspension systems.

Small cars for transportation of people is probably the area where suspension systems have been developed to the highest standard. The suspension system should make it comfortable to sit in the car both for the driver and the passengers. The vehicle should be easy and safe to drive on the road. Certain angles in the suspension identified as caster, camber, toe in/out, and others, control the vehicle's behavior on the road. The McPherson strut and

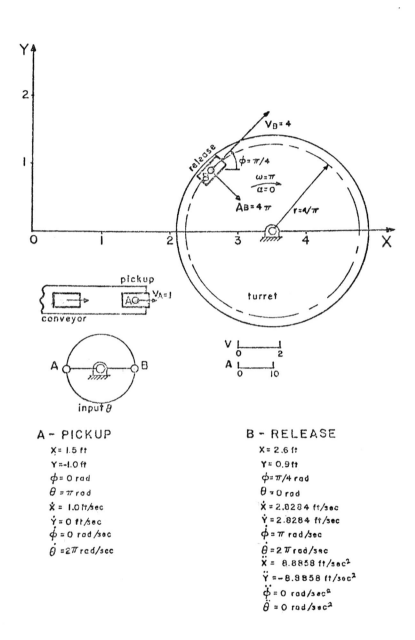

FIGURE 8.10: Synthesis problem specifications for MECSYN [33]

TABLE 8.4

Input data for the Stephenson 3 mechanism.

Index	Name	X-pos	Y-pos	Z-pos
0	"P0"	0.2578	0.01	0.0
1	"P1"	0.2661	0.6337	0.0
2	"P2"	0.11	0.6864	0.0
3	"P3"	0.1085	0.4935	-0.06
4	"P4"	0.4849	0.1204	-0.06
5	"P5"	0.3959	-0.1262	0.0
6	"P6"	0.7925	0.2743	0.0
7	"P7"	0.2578	-0.5	0.0

Point	Type	Link-incidence	Joint-direction
0	"revolute"	(nil 0)	(0 0 1)
1	"revolute"	(1 nil)	(0 0 1)
2	"revolute"	(1 3)	(0 0 1)
3	"revolute"	(2 nil)	(0 0 1)
4	"revolute"	(2 4)	(0 0 1)
5	"revolute"	(4 0)	(0 0 1)
6	"revolute"	(4 3)	(0 0 1)

Name	Link	Member	Cross-section	Dimensions
"Links"	default	default	"circular"	(0.04 0.04)

Type	Point	Direction	Magnitude	Loaded-link
Torque	0	(0.0 0.0 -1.0)	7.5	0

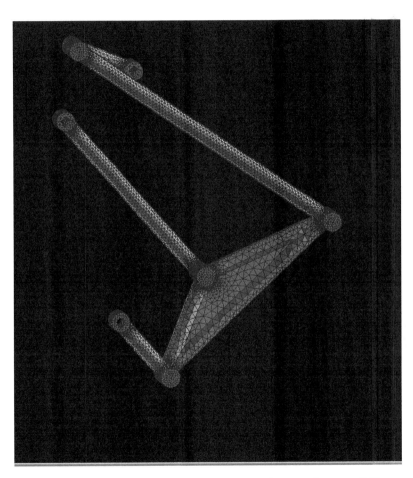

FIGURE 8.11: Meshed Stephenson 3 six-bar mechanism [33]

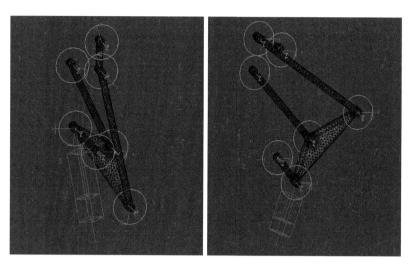

FIGURE 8.12: Stephenson 3 six-bar mechanism in pickup and drop position [33]

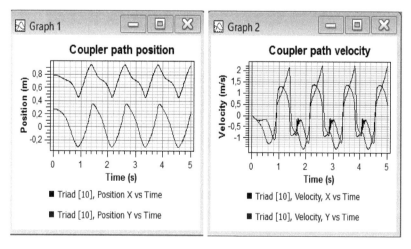

FIGURE 8.13: Stephenson 3 coupler position and velocity curves [33]

the double wishbone are two well-known examples of suspension systems for passenger cars. The axes that the front wheel rotates about when steering is also a key entity when working to improve the properties of the car's suspension; for more details refer to standard books on car suspensions.

8.2.1 Double wishbone suspension

A wishbone refers to a y-shaped bone in birds and some other animals. Two of the links used in one of the most popular front suspensions for personal cars is named after his bone, i.e., the double wishbone suspension. Both these links (wishbones) are connected to the car body with two revolute joints with common axis. In these revolute joints there are usually rubber bushings to isolate the vibrations and shocks from the suspension from the car to minimize rough motion of the car body. The knuckle, the link where the wheel is mounted, is connected to the lower and upper wishbone links with ball joints. On one side of the knuckle there is another ball joint for connecting to the steering rod. Push/pull by the steering rod will cause the knuckle and the wheel to rotate around the ball joints on the lower and upper wishbone arms and thus control the steering motion of the wheels. This rotation axes for the wheel assembly is critical for the road handling of the car.

The angle between this steering axes and a vertical line seen from the side is called the caster angle, which can be positive or negative. The angle between the steering axes and a vertical line seen from the front is called the camber angle. The angle for how much the wheels are spreading out is called toe-out or toe-in if the wheels are plowing. Extending the steering axes to the road is another important measure; if this point on the road level is inside or outside the center of the tire (where the tire transmits the forces to the road). There are other parameters used to characterize a front suspension, but these are the most important variables. It is also important that the wheels are keeping the direction stable when it hits a bump in the road and that it does not move too much in and out hitting the bump. The longer the wishbone arms are the less in-out motion you will get from a bump in the road, but there will be space limitation in the car body and a compromise has to be found.

Figure 8.14 shows a double wishbone suspension modeled and meshed with solid mesh in RaMMS and Figure 8.15 shows the same suspension transferred into FEDEM. The input files for coordinates, constraints and geometry shapes for the RaMMS application are shown in Table 8.5; refer also to Tables 4.1, 4.2 and 4.3 in Chapter 4 for discussions around these data inputs. This 3D model of a suspension system needs more modeling features than the planar mechanisms presented earlier. The lines labeled index 0 – 9 (see first section in Table 8.5) represent coordinates for positioning entities like joints and endpoints for springs and dampers as we have used earlier for planar mechanisms. In the second data section, the joint topology is defined in the same way as shown earlier, however, note how the joint-directions are specified for this spatial mechanism.

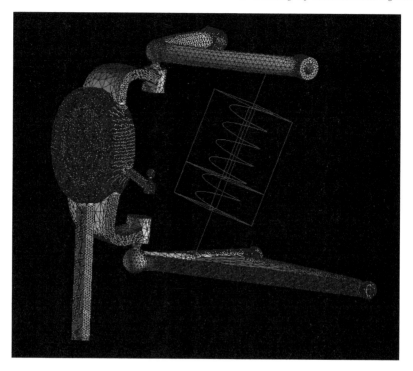

FIGURE 8.14: Meshed double wishbone suspension

In the third data section in Table 8.5 the link shapes are defined in the same way as before with name, link reference, member reference, cross section type and cross-section dimensions, however, more input is needed to specify the more complex 3D shapes of the links for this suspension system. For the member representing the upper part of the steering-knuckle two extra control points (10 11) are needed with corresponding weights (0.1 0.9) for controlling this member's spline path. The corresponding control points and weights for the steering knuckle lower part is (12 13) and with weights (0.1 0.9), respectively. The steering knuckle member connecting to the steering ball needs one control point (14) with corresponding weight (0.4). In the link shape data in Table 8.5 we can also see that one of the members for the lower suspension arm has specified a spline with three control points (15 16 17) with corresponding weights (0.7 0.7 0.1). The upper suspension arm has specified two members as spline curves with control points (18 19 20) and (21 22 23) with corresponding weights (0.5 0.5 0.5) for both splines. This input has been quite demanding to enter manually, but the link editor presented in Figure 7.6 in Section 7.1.2 can simplify this work substantially.

For space limitations in Table 8.5 the suspension spring and damper details and loading are not included.

TABLE 8.5
Input data for the double wishbone Suspension.

Index	Name	X-pos	Y-pos	Z-pos
0	"upper-ball"	0.0	-0.15	0.3
1	"upper-front"	-0.085	-0.02	0.3
2	"upper-back"	0.085	-0.02	0.3
3	"lower-front"	-0.06	0.04	0.08
4	"lower-back"	0.17	0.04	0.08
5	"lower-ball"	0.0	-0.15	0.08
6	"steering-ball"	0.085	-0.20	0.185
7	"spindel"	0.003	-0.228	0.2
8	"sprDmp-bottom"	0.006	-0.08	0.082
9	"sprDmp-top"	0.0	0.0	0.3
10	"upper-knuckle-nurb1"	0.0	-0.23	0.3
11	"upper-knuckle-nurb2"	0.0	-0.215	0.3
12	"lower-knuckle-nurb1"	0.0	-0.23	0.08
13	"lower-knuckle-nurb2"	0.0	-0.215	0.08
14	"steering-link-nurb"	0.07	-0.23	0.19
15	"lower-m0-nurb1"	0.0	-0.05	0.08
16	"lower-m0-nurb2"	0.05	0.04	0.08
17	"lower-m0-nurb3"	0.12	0.038	0.08
18	"upper-m0-nurm1"	-0.02	-0.14	0.3
19	"upper-m0-nurm2"	-0.05	-0.1	0.3
20	"upper-m0-nurm3"	-0.082	-0.05	0.3
21	"upper-m1-nurm1"	0.02	-0.14	0.3
22	"upper-m1-nurm2"	0.05	-0.1	0.3
23	"upper-m1-nurm3"	0.083	-0.05	0.3
24	"rim-connection"	0.03	-0.23	0.0

Point	Type	Link-incidence	Joint-direction
0	"ball"	(0 2)	(0 0 1)
1	"revolute"	(nil 2)	(-1 0 0)
2	"revolute"	(nil 2)	(-1 0 0)
3	"revolute"	(1 nil)	(-1 0 0)
4	"revolute"	(1 nil)	(1 0 0)
5	"ball"	(0 1)	(0 0 -1)
6	"ball"	(0 nil)	(0 0 1)
7	"revolute"	(0 3)	(0 -1 0)
8	"free"	(1 nil)	(0 0 -1)
24	"free"	(3 nil)	(0 0 1)

Name	Lnk	Mem	Cross-section	Dimensions	Points	Weights
"knuckle"	0	0	"nil"			
"knuckle"	0	1	"nil"			
"knuckle"	0	2	"rectangular"	(0.02 0.01 0.09 0.01)	(10 11)	(.1 .9)
"knuckle"	0	3	"nil"			
"knuckle"	0	4	"rectangular"	(0.02 0.01 0.1 0.011)	(12 13)	(.1 .9)
"knuckle"	0	5	"circular"	(0.01 0.01)	(14)	(.4)
"lower-arm"	1	0	"rectangular"	(0.015 0.005)	(15 16 17)	(.7 .7 .1)
"lower-arm"	1	1	"rectangular"	(0.015 0.005)		
"lower-arm"	1	2	"nil"			
"lower-arm"	1	3	"rectangular"	(0.015 0.005)		
"lower-arm"	1	4	"nil"			
"lower-arm"	1	5	"circular"	(.022 .022 .015 .015)		
"upper-arm"	2	0	"circular"	(0.02 0.02)	(18 19 20)	(.5 .5 .5)
"upper-arm"	2	1	"circular"	(0.02 0.02)	(21 22 23)	(.5 .5 .5)
"upper-arm"	2	2	"nil"			
"rim-con"	3	0	"rectangular"	(0.02 0.01)		

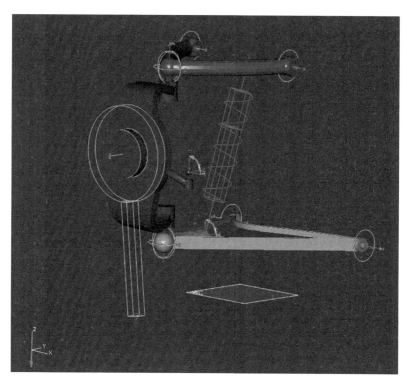

FIGURE 8.15: Double wishbone suspension in FEDEM

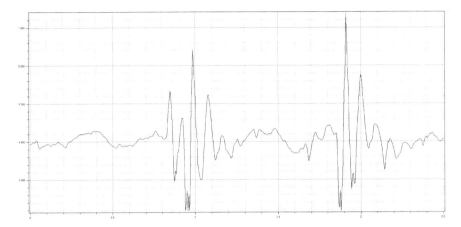

FIGURE 8.16: Road excitation input to the FEDEM simulation

For this pilot implementation of the RaMMS KBE application, some additional data is imported into FEDEM. A load profile from road excitation in Figure 8.16 and a function controlling the steering motion is shown in Figure 8.17. A simulation time of 2.5 seconds with time steps of 0.002 seconds has been selected. A typical stress result is shown in Figure 8.18. The stresses are only displayed for the steering knuckle. Some snapshots from result stress animation are shown in Figure 8.19 for the times 0.2, 0.6, 1.0, 1.4, 1.8 and 2.2 seconds, respectively. This simulation combines vertical motion from bumps in the road and steering motion; refer to Figures 8.16 and 8.17. This simulation produces forces in the suspension spring and damper; see Figure 8.20.

As you can see with just about one page of input data (see Table 8.5) almost all information needed to generate the quite demanding simulation input for the double wishbone simulation is generated. This includes link geometries connected with joints with quite detailed geometries. Based on these link geometries, FE meshes are generated with the mesh density adjusted for blended sharp edges. RBE2 connections (see Section 3.7.1) are generated at each joint ready for direct joint modeling in FEDEM. For 3D geometries, setting up the input data could be somewhat demanding, however, with the interactive input tools presented in Chapter 7 the modeling of these data will be significantly simplified. The effectiveness gains are quite extensive if you are building just one version of the geometries, however, if you need to develop more versions of the geometries with corresponding FE meshing the work will only be, say, less than 10% of the work if you should do this manually in the usual way.

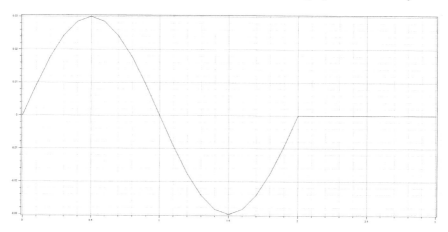

FIGURE 8.17: Steering input to the FEDEM simulation

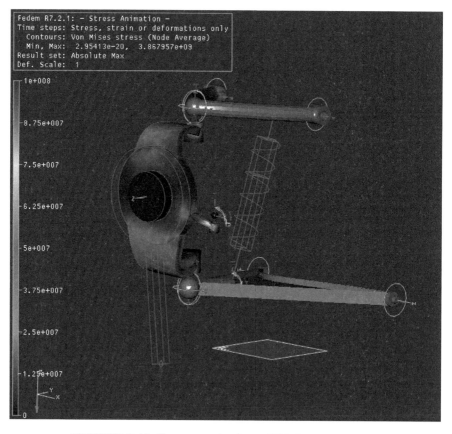

FIGURE 8.18: Stress results from FEDEM simulation

FIGURE 8.19: Suspension stress animation from FEDEM simulation. Sequence top to bottom, left to right

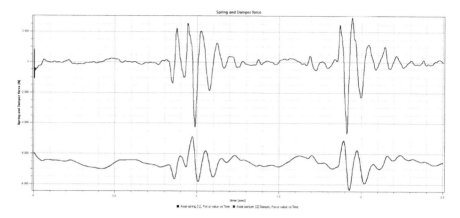

FIGURE 8.20: Spring and damper force from the FEDEM simulation

8.3 Cranes and Robots

Cranes and robots are a very large area for application of mechanisms. You will find cranes on trucks, boats, building sites and on docks all over the world. Also on offshore installations there are cranes in all sizes. Some are purely manually controlled, and some are actively controlled with heave compensation for avoiding a load crashing into the deck of a boat moving in the waves. There is a somewhat gradual transition between what could be called a crane and what could be called a robot. Robots are probably the largest area for applications of mechanisms nowadays. In production facilities, robots are replacing manual work like welding, riveting, assembling and handling of goods. In this section we are using a scaled offshore crane model as an example for this type of mechanisms. These types of mechanisms distinguish themselves from linkages and suspension systems in previous sections with no, or few, kinematic loops and the joint positions as design variables are not as natural a choice. This is because these structures are usually tree like structures that could be built from the ground/foundation and out. For these kinds of mechanisms, the design object could be the individual parts, and when the parts are designed they could be positioned following each other to assemble the mechanism. Using the joint coordinates as design variables is not the only choice here, but we will demonstrate a crane example where this still works very well.

Typical for cranes and robots, there could be more than one joint between the same two links. This is not common (or possible) in classical mechanism simulation with rigid bodies. For instance a door will usually be connected to the wall with two or more hinges, but for rigid body simulation of the door you model this as one single revolute joint because using two or more revolute joints will make the model over-constrained. However, in the simulation tool referred to in his book [3], the links are modeled as flexible parts using a finite element formulation, and over-constraining does not usually pose a problem. Back to the door example—you could model the door constraints with more than one revolute joint, for instance one for each hinge. If the rotational axes for the different hinges do not coincide exactly the door can still move when the door link deforms.

8.3.1 Offshore crane

The coordinates for the crane example here are from an offshore crane that was down-scaled for laboratory tests. Two large brackets are mounted on a table that can turn around a revolute joint with vertical joint axes. The crane is simplified using only three more links: two middle arms on top of the brackets and an outer arm on top of the middle arms. Hydraulic cylinders between the turning table and the middle arms, and between the middle arms and the outer arm, are modeled as axial springs; see Figure 8.21. For this model, a surface shell mesh is used. Linearly varying rectangular cross section profiles are used as connections between joints, as shown in Table 8.6; refer also to Tables 4.1, 4.2 and 4.3, and others in Chapter 4. As can be seen in Table 8.6, quite a few connections have been removed (marked nil) for the turning table (slewingplate). This mechanism has three types of joints: revolute, free and fixed. The revolute joints are for constrained rotations, and the free joint for other points of interest in the mechanism as for instance connecting points for springs. What are called fixed joints (also named rigid joints) are used to connect the lower arms (brackets) rigidly to the turning table (slewingplate).

This model in our AML application (RaMMS) is transferred as simulation input to FEDEM; see Figure 8.22. Some simulation modeling details of the FE meshed model in FEDEM are shown in Figure 8.23. Surface shell elements are generated for this model and in each joint position rigid elements (RBE2) are generated for the meshes to be ready for assembling with joints.

A combined FEDEM simulation with swing and lift has been performed on this model: lifting and moving loads by rotating the slewingplate through its vertical revolute joint (simulating an electric or hydraulic motor), and controlling the length of the axial springs for lifting or lowering the crane (simulating a prescribed handling operation of the crane). The simulation is used for dimensioning the hydraulic or electric system driving the crane, but also for dimensioning links and arms of the crane both with respect to deflections and material stresses. The extra input to run the different simulations is easily edited interactively in the FEDEM system. The simulation time was set to 10 seconds with integration time steps of 0.001 seconds. The modeled input functions for swing and actuator motion are shown in Figure 8.24.

An example stress plot from a FEDEM simulation is shown in Figure 8.25. Twelve stress animation snap shots are shown in Figure 8.26.

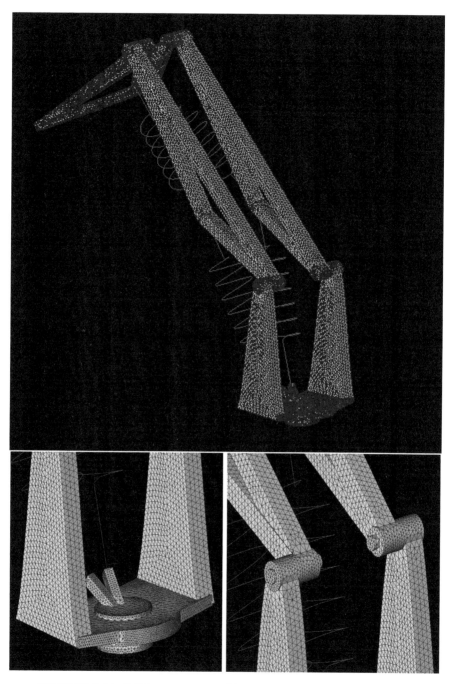

FIGURE 8.21: Offshore crane in AML with some FE mesh details

TABLE 8.6
Input data for the offshore crane.

Index	Name	X-pos	Y-pos	Z-pos
0	"rotation"	-0.80	-1.85	0.62
1	"spr2-bot"	-0.83	-1.95	0.73
2	"spr1-bot"	-0.77	-1.95	0.73
3	"center-1"	-0.60	-1.89	0.66
4	"center-2"	-1.00	-1.89	0.66
5	"revoluteA"	-0.60	-1.85	1.22
6	"revoluteB"	-1.00	-1.85	1.22
7	"spr1-top"	-0.65	-2.28	1.45
8	"spr2-top"	-0.95	-2.28	1.45
9	"revoluteC"	-0.72	-2.73	2.11
10	"revoluteD"	-0.88	-2.73	2.11
11	"spr4-top"	-0.85	-2.82	1.91
12	"spr3-top"	-0.75	-2.82	1.91
13	"tipp"	-0.80	-3.5	1.63

Point	Type	Link-incidence	Joint-direction
0	"revolute"	(0 nil)	(0 0 -1)
1	"free"	(0 nil)	(1 0 0)
2	"free"	(0 nil)	(1 0 0)
3	"fixed"	(0 1)	(1 0 0)
4	"fixed"	(0 2)	(-1 0 0)
5	"revolute"	(3 1)	(1 0 0)
6	"revolute"	(4 2)	(-1 0 0)
7	"free"	(3 nil)	(0 1 0)
8	"free"	(4 nil)	(0 1 0)
9	"revolute"	(3 5)	(-1 0 0)
10	"revolute"	(4 5)	(1 0 0)
11	"free"	(5 nil)	(0 1 0)
12	"free"	(5 nil)	(0 1 0)
13	"free"	(5 nil)	(1 0 0)

Name	Link	Member	Cross-section	Dimensions Points-list
"SlewingPlate"	0	0	"rectangular"	(0.025 0.025 0.025 0.025)
"SlewingPlate"	0	1	"rectangular"	(0.025 0.025 0.025 0.025)
"SlewingPlate"	0	2	"rectangular"	(0.25 0.03 0.25 0.025)
"SlewingPlate"	0	3	"rectangular"	(0.25 0.03 0.25 0.025)
"SlewingPlate"	0	4	"nil"	
"SlewingPlate"	0	5	"nil"	
"SlewingPlate"	0	6	"nil"	
"SlewingPlate"	0	7	"nil"	
"SlewingPlate"	0	8	"nil"	
"SlewingPlate"	0	9	"nil"	
"LowerArm1"	1	0	"rectangular"	(0.25 0.055 0.055 0.055)
"LowerArm2"	2	0	"rectangular"	(0.25 0.055 0.055 0.055)
"MiddleArm1"	3	default	"rectangular"	(0.055 0.055 0.055 0.055)
"MiddleArm2"	4	default	"rectangular"	(0.055 0.055 0.055 0.055)
"UpperArm"	5	0	"nil"	
"UpperArm"	5	7	"nil"	
"UpperArm"	5	default	"rectangular"	(0.035 0.035 0.035 0.035)

Type	Point-from	Point-to	Incident-links	Stiffness/Damping
"spring"	2	7	(0 3)	1000000
"spring"	1	8	(0 4)	1000000
"spring"	7	12	(3 5)	1000000
"spring"	8	11	(4 5)	1000000

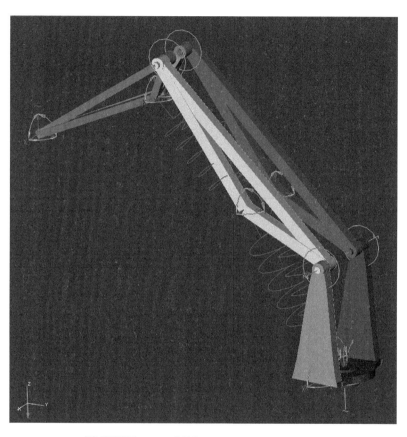

FIGURE 8.22: Offshore crane in FEDEM

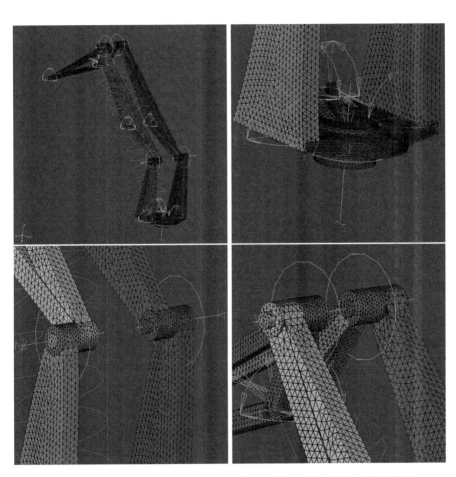

FIGURE 8.23: Offshore crane with modeling details in FEDEM

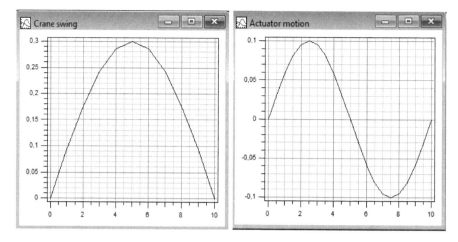

FIGURE 8.24: Offshore crane input swing and actuator motion

8.4   Wind Turbines

The last type of mechanism we will discuss in this text is wind turbines. A large variation of wind turbines is popping up onshore and offshore with wind turbine blades as long as 60–80 meters, that is with a total wingspan of up to 160 meters. Feasibility simulation of such extreme structures is very important and being able to quickly generate simplified geometries for these kinds of structures that could withstand these extreme dynamic forces is crucial, both for production in optimal wind conditions, but also to withstand the wind loading for extreme wind conditions.

8.4.1   An offshore wind turbine model

An example of offshore reference wind turbine is presented in [17]. This is designed as a 10 MW offshore wind turbine with a wing span (diameter) of 140 meters. It is designed for a wing tip speed of 90 meters per second, that is, about 5 seconds per revolution. A simplified simulation model is generated in our AML application; see Figure 8.27. The detailed wing profile is not taken into consideration; only the stiffening structure called the spar has been modeled. A close-up of the same model with FE surface shell mesh is shown in Figure 8.28. The input data for this model is shown in Table 8.7. Linearly varying rectangular profiles are used to model the connections between the joints (the tower for the wind turbine is not included in the model).

The wind turbine model is imported into FEDEM and shown in the FE-DEM graphic user-interface in Figure 8.29. A close-up with FE shell mesh

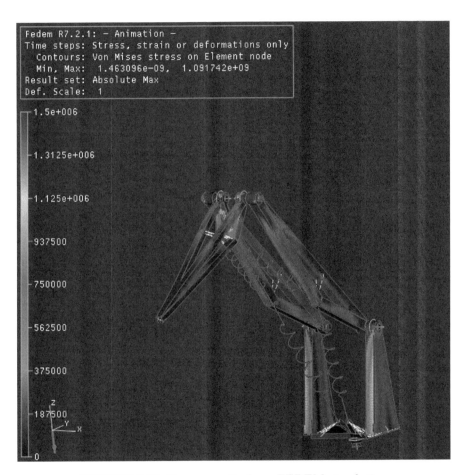

FIGURE 8.25: Stress results from FEDEM simulation

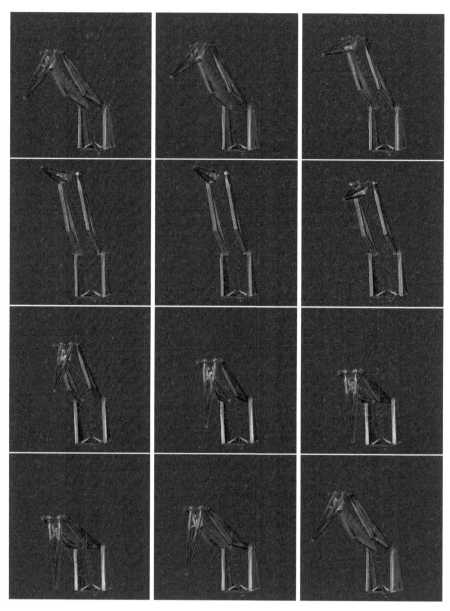

FIGURE 8.26: Offshore crane stress animation from FEDEM simulation. Sequence top to bottom, left to right.

FIGURE 8.27: Wind turbine in AML

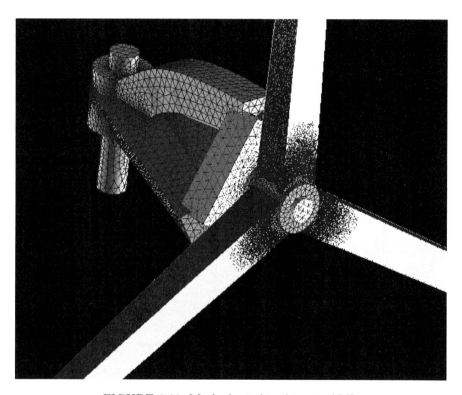

FIGURE 8.28: Meshed wind turbine in AML

TABLE 8.7
Input data for the wind turbine.

Index	Name	X-pos	Y-pos	Z-pos
0	"Crank-bearing"	0.0	0.0	0.0
1	"Blade-1"	0.0	0.0	69.85
2	"Blade-2"	0.0	-60.4919	-34.925
3	"Blade-3"	0.0	60.4919	-34.925
4	"con-1"	-4.0	0.0	5.20
5	"con-2"	-4.0	-4.5033	-2.6
6	"con-3"	-4.0	4.5033	-2.6
7	"Top1"	-20.0	0.0	-4.8

Point	Type	Link-incidence	Joint-direction
0	"revolute"	(1 0)	(1 0 0)
1	"free"	(0 nil)	(1 0 0)
2	"free"	(0 nil)	(1 0 0)
3	"free"	(0 nil)	(1 0 0)
4	"fixed"	(1 2)	(1 0 0)
5	"fixed"	(1 2)	(1 0 0)
6	"fixed"	(1 2)	(1 0 0)
7	"revolute"	(2 nil)	(0 0 -1)

Name	Link	Member	Cross-section	Dimensions
"Crank-link"	0	0	"rectangular"	(3.37 3.37 1.62 0.29)
"Crank-link"	0	1	"rectangular"	(3.37 3.37 1.62 0.29)
"Crank-link"	0	2	"rectangular"	(3.37 3.37 1.62 0.29)
"Crank-link"	0	3	"nil"	
"Crank-link"	0	4	"nil"	
"Crank-link"	0	5	"nil"	
"Connection1"	1	default	"rectangular"	(3.0 3.0 3.0 3.0)
"connection2"	2	0	"nil"	
"connection2"	2	1	"nil"	
"connection2"	2	3	"nil"	
"Connection2"	2	default	"rectangular"	(3.0 3.0 4.0 4.0)

Type	Point	Direction	Magnitude	Loaded-link
Torque	0	(-1.0 0.0 0.0)	1000000.0	0

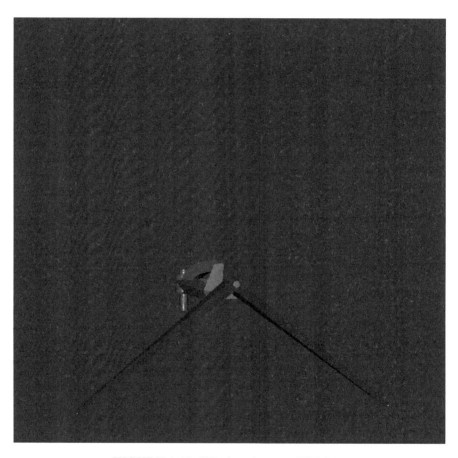

FIGURE 8.29: Wind turbine in FEDEM

FIGURE 8.30: Meshed wind turbine in FEDEM

is shown in Figure 8.30. Different types of dynamic simulations of the wind turbine can be carried out in the FEDEM system with some additional input through the FEDEM use interface as for instance wind loading on the turbine blades and the control and gear operation. To verify the model in FEDEM, the links are reduced and a simulation is carried out by turning the turbine blades a couple of rotations by introducing a large torque in the center revolute joint of the propeller. To carry out realistic simulation in FEDEM, wind forces on the propeller have to be calculated and added into the model. In addition, a control system for the turbine has to be implemented in the FEDEM simulation model. Also the gear transmission and the turbine tower need to be included to make a realistic simulation model. The main purpose of this modeling example is to demonstrate that these types of structures can also be generated in our KBE application and imported into FEDEM as simulation input.

This wind turbine example also demonstrates some limitations in the current version of the KBE pilot application RaMMS. For instance we did not model the turbine tower because the modeling features implemented in the current version of the KBE pilot do not easily adjust to a case where a revolute joint connects to a structure axially. Some additional elements need to be added to the KBE pilot, for instance an axle element. Limitations in the current implementation of the KBE pilot will be discussed in the next chapter.

8.5 Optimization for Automation in Mechanism Design

This section is based on the master's thesis of Arnt Underhaug Lima [25] and demonstrates with how optimization algorithms can be used to improve design performance using dynamic simulation to comply with specified performance criteria.

8.5.1 Optimization approaches demonstrated on a four-bar mechanism

In this case, a relatively simple mechanism, with relatively simple analysis and parameterization, is optimized. Specifically, the location of a joint is varied, with the desired path of a point on the coupler curve as the objective. The main goal is to demonstrate the optimization system functions as a whole, with all of the components, as well as demonstrating the complexity these components bring. The geometry for the four-bar referred to is shown in Figure 8.31.

For each point in the design space, a procedure for analyzing the performance of the design is required. The objective is to find the mechanism that has the straight line behavior. As the path traced by the tip of the coupler is the chief concern, the path of this point constitutes the primary input of the analysis. In addition to the path, the analysis requires two points that define a line segment which forms the reference path the mechanism should follow.

The first step in the analysis, is to find the start and end time. These values define the time interval that should be used when comparing the mechanism path to the reference path. Since the inertia and stiffness of the mechanism are different for differing design points, the start and end time cannot be hard-coded; they need to be detected. Hence the analysis detects when the mechanism is closest to the first point and selects this time as the start time. Similarly, the end time is defined by when the mechanism is closest to the second point.

The path of the coupler is defined with time as a parameter, whereas the reference path, and the goal of creating a straight line, are both time invariant. Consequently the coupler path should be made time invariant by using a new parameterization, similar to how the substitution $ds = v(t)dt$ is used to solve line integrals in calculus. For convenience, the path is defined to be $r(t(\tau))$ with the properties $t(\tau = 0) = t_{start}$, $t(\tau = 1) = t_{end}$ and $dr = d\tau = Const$.

By creating a path for the reference line, also parameterized with τ, the distance between the paths is easily calculated as $D = r(\tau) - r_{ref}(\tau)$. This distance may be described as a function $D(\tau)$. Integrating the L_2 norm of this function, a measure of how well the coupler follows the reference path is obtained. For the actual implementation of this, all operations are written as vectorized expressions, and the integral $\int_0^1 |D(\tau)|^2 d\tau$ is solved numerically with the trapezoidal rule. Further, the property $dr/d\tau = Const$ ensures that the formulation is not affected if, for instance, the mechanism has a large distance to the reference path, for a short time at significant speed.

For the potential use as a constraint, a measure of how much of the path of the coupler is above the reference path is needed. In other words, the constraint is violated if the y component of $D(\tau)$ is positive, and a measure of the degree of violation is required (the y direction is directed up in Figure 8.31). The measure of the constraint is obtained with an integral similar to the previous integral, only a slight alteration of the function $D(\tau)$ is required. The aforementioned alteration is obtained by taking the value of the new function to be 0 if the y component of $D(\tau)$ is negative, else-wise the original value of $D(\tau)$ is used. The following optimization approaches will be tested for this case:

- Optimization using the model and finite difference sensitivities, with the constraint ignored

- Optimization using the model and finite difference sensitivities, with the constraint

- Optimization on a surrogate model, with the constraint included

The SLSQP algorithm was used for all the executions mentioned above; see

Section 5.7. This optimization problem is formulated in the YAML [15] format as input to the optimization tool. This includes both a redefinition of the mechanism input for the KBE tool and the input to control the optimization including definition of design variables, the objective function, and the constraint. The syntax for this input will not be discussed here.

8.5.1.1   Optimization without constraints

The first execution, with no constraint, converged to a point a few millimeters from the known solution, with a slightly improved performance. As the difference in both the design space point, and performance, were very small, the point that the algorithm converged to can be thought of as the same point as the known, working solution. Nevertheless, an actual improvement to the analytic design was found.

Figure 8.32 shows the designs that were evaluated by the optimization algorithm, for the execution without a constraint. A circle indicates that the two values corresponding to this point were used to generate a design, in an iteration of the optimization algorithm. Since the two design parameters happened to be the planar coordinates of a physical point, each circle has a physical interpretation as the location of Point 2 in Figure 8.31, and the two figures use the same coordinate system.

Scatterplots are used throughout this section. They are used to visualize the points evaluated by the optimization system. Hence some observations on how these plots are used are in place. From the optimization algorithms perspective, every point in the design space is a dimensionless vector of real numbers. This vector might correspond to spatial coordinates, and thus it might have a physical interpretation, but this is not a necessity. The scatterplots are used to illustrate the behavior of algorithms; consequently, the physical interpretation is of less importance, e.g., the clustering in Figure 8.32 is more interesting than the coordinates of the convergence point. Since any proper naming of the axes would be rooted in the spatial interpretation, and the spatial interpretation is of lesser interest, all naming is omitted.

From Figure 8.32 it is possible to see that the algorithm fairly quickly goes from the lower left corner, where it started, to a location near the convergence point. In some later iteration the algorithm jumps to the upper left corner, and again quickly moves to the convergence point. This last step repeats, multiple times, and the quick movement appears to be happening along a line.

8.5.1.2   Optimization with constraint

When the constraint was introduced, the problem converged to a markedly different point, most notably the design variable corresponding to the y value became negative, and the value of the objective increased slightly. In order to confirm that the problem formulation was correct, both the known reference design, and the new solution were analyzed directly in FEDEM, without re-

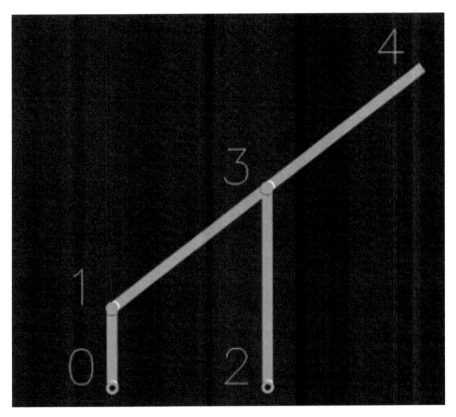

FIGURE 8.31: Geometry of straight line generator [25]

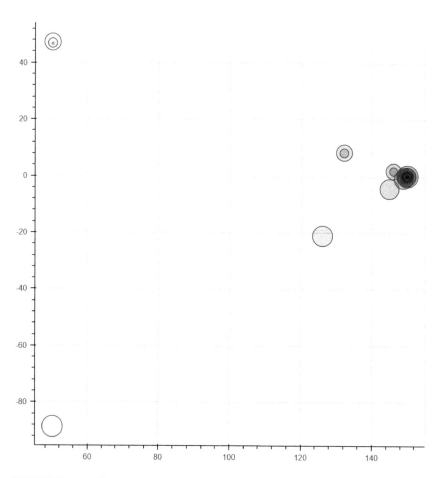

FIGURE 8.32: Design parameter variation during optimization without constraints. (Light gray designates high cost, and dark gray low cost; size of circle indicates iteration, with diameter decreasing for each iteration) [25]

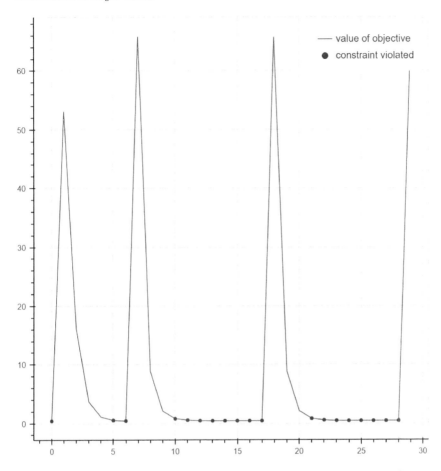

FIGURE 8.33: Value of cost function, for each iteration [25]

lying on the optimization system. As required by the constraint, the FEDEM analyses confirms that the path never obtains a negative y value.

The behavior of the optimization algorithm is shown in Figure 8.33. In this figure, one can note that the algorithm rapidly decreases the objective function, but then is blocked by the constraint. Soon the algorithm does a jump and again rapidly decreases the objective; this behavior is similar to the behavior for the unconstrained case.

8.5.1.3 Optimization on surrogate model

For the Design of Experiment(DoE), 40 cases were evaluated. The results of these experiments are visualized in Figure 8.34, which has the same logic and interpretation for the placement of the circles corresponding to Figure

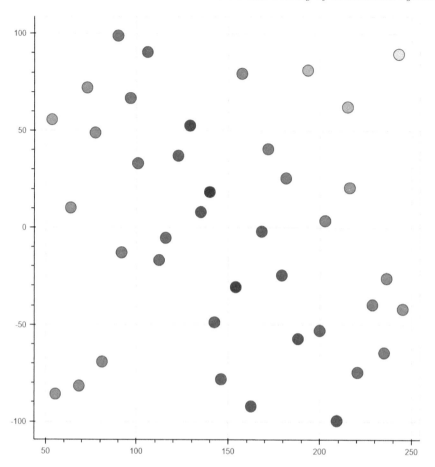

FIGURE 8.34: Samples in the design of experiments (Light grey denotes high cost, and dark grey low cost) [25]

8.32. From these observations two surrogate models were constructed using OpenMDAO: One surrogate model (SM) for modeling the objective function, and one for modeling the constraint. When the optimization algorithm was used on these functions, the algorithm converged to a design practically equal to the design found when the algorithm worked on the KBE model directly.

In order to confirm the accuracy of the approximations, the surrogate model was used to create approximations for the value of the objective function in specific points. Such approximations were set up for every point evaluated when the constrained case was solved by direct optimization.

8.5.1.4 Discussion of case results

The effectiveness of the somewhat abstract formulations of the objective and the constraint also supports an optimistic view of what optimization can achieve in the design of mechanisms. Specifically, the results show that when optimization is used, designs that already are known can be rediscovered. When compared with traditional synthesis methods, the rediscovery process might even involve less work, if the optimization system already has been created, since only a declaration of the objective is required.

The solution to the constrained problem is obtained with a simple extension of the formulation for the unconstrained case. The ease of extending the unconstrained case offers some insight into the generality and abstractness that naturally comes when specifying design problems as optimization problems. Thus, in a work-flow using optimization, the requirements can be iteratively refined.

In addition to simplifying the work-flow, the formulation of design problems as optimization problems may provide better designs, when compared with traditional analytic methods. The unconstrained case demonstrated this phenomenon. The performance, as measured, was worse for the reference design than for the design found by the optimization algorithm. This can be explained by the fact that the optimization process incorporates more physical phenomena, such as inertia and flexibility. Thus one may consider the unconstrained optimization problem to be a better description of the design problem, as the analytic specification is not able to provide a design that has the same performance.

Increases in the objective function, as seen in both the constrained and unconstrained case, can be attributed to the line search procedure used in SQP algorithms [34].

Since finite difference was used on both the constrained and unconstrained case, the issue of setting the step size had to be met. Since the design parameters are physical parameters, any engineer with an intuition for the amount of change required to obtain a meaningful, but small, change in the design, can come up with a sensible step size. Using this method the step size was set to 10.0 mm, which was used when both cases converged.

8.5.2 Optimization approaches demonstrated on a Stephenson 3 mechanism

In this case, a more complex optimization problem is explored. When the additional complexity is introduced, the case becomes more informative, and can better express the suitability of optimization in the mechanism design process. The effectiveness of the current system can also be more deeply investigated.

Myklebust's PhD thesis [33] has a description of a design case which has been repurposed in our case, as presented in Section 8.1.2. The mechanism has the purpose of taking parcels from a conveyor belt and placing them on a rotating turret. A Stephenson 3 mechanism is used to solve the problem. Figure 8.10 shows the parameters Myklebust used to define the problem, and many of the parameters can be considered state variables when the problem is reformulated as an optimization problem. Some of the variables, such as the positions, will be driven by the optimization algorithm since they are required to instantiate the KBE model; thus they form the basis for the design variables. Variables such as the velocities can only be obtained from FEDEM, and thus they should be considered state variables. They are therefore used when formulating the objective and constraints.

The path is available in FEDEM when a design has been evaluated. Myklebust uses numerical values only for the pickup and deposit points, as the system used requires the specifications to be given for specific positions of the mechanism. Since the complete path is available in FEDEM, a more complex case could be developed, where for instance the maximum acceleration for the whole path is a constraint. Such adaptations of the case have not been attempted, and the specifications of the original case are used directly.

The original case solves the design problem exactly, as a consequence of the design tooling used. When the same case is to be reformulated as an optimization problem, it is not necessarily known what the most appropriate formulations of the objective and constraints are. For this reason, it was decided to do a design of experiments (DoE), storing the state variables that were used in the original case as the results of each experiment. With such a setup, many formulations of the objective can be created after all the expensive simulations have been completed.

As stated, most of the design parameters, in the original case, are associated with either the pickup point or the deposit point. These responses are easily obtained as time series from FEDEM. However, the design problem requires the responses for specific positions of the coupler point. Consequently, the times for the pickup and deposit action have to be calculated, and this is done by taking the two time steps for which the coupler is closest to the pickup point and deposit point, respectively.

8.5.2.1 Parameterization

By making observations on the jamming behavior for this mechanism, the causes for this behavior can be recognized. Taking all point numbers from Figure 8.35, one might note that, by only considering the link segments $(3; 4)$ and $(4; 5)$, and considering the joint in Point 3 to be fixed, all possible locations of Point 5 are easily defined. These locations form a disc, represented by the white area in Figure 8.36 , and the area has to contain the circle defining the path of the crank (i.e., segment $(0; 5)$). The d_i values can be thought of as some sort of clearance size values. They have to be positive for the path of the Point 5 on the crank to be in the area Point 5 can reach as part of the segments $(3; 4)$ and $(4; 5)$.

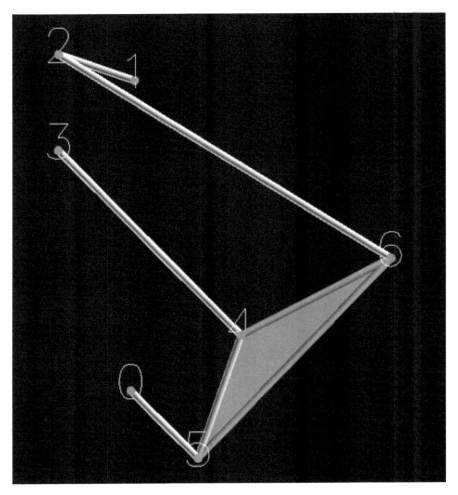

FIGURE 8.35: The geometry of the Stephenson 3 mechanism, generated by
the KBE system, in the form that solves the original case [25]

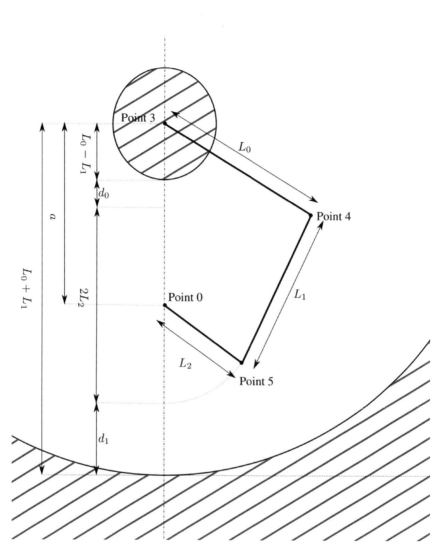

FIGURE 8.36: Illustration of the area point 5 cannot reach, with the relevant lengths shown [25]

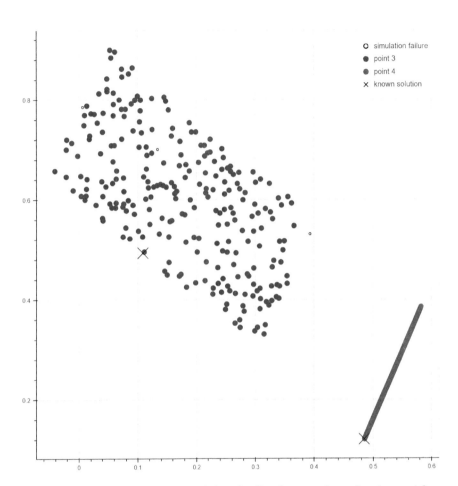

FIGURE 8.37: Designs evaluated for the Stephenson 3 mechanism, with new parameterization (two points are plotted for each design evaluated, with the position of each point corresponding to the points position in the global reference system) [25]

It is possible to obtain additional insight from Figure 8.36, by noting that the relation below must necessarily hold.

$$L_0 + L_1 = (L_0 - L_1) + d_1 + 2L_2 + d_0 \tag{8.1}$$

$$L_1 = (d_1 + d_0)/2 + L_2 \tag{8.2}$$

By solving for L_1, it can be shown that this variable is a function of d_0, d_1 and L_2. Thus L_1 can be removed as a parameter of the mechanism; it can be derived from other values. Further, as the value d_i does not affect the geometry directly, the sum can be rewritten as $d_0 + d_1 = D$, with the move limit $D > 0$.

$$L_1 = D/2 + L_2 \tag{8.3}$$

By using the relation in Equation 8.3, the mechanism can be parameterized such that the jamming behavior can be eliminated. First, the mechanism is defined with a set of lengths and the angles of the joints (in the design position). Then Equation 8.3 is substituted for L_1. From such a description, the points that serve as input to the KBE system can be calculated in a fashion similar to the transformation from polar to Cartesian coordinates.

This optimization problem is formulated in the YAML [15] format as input to the optimization tool. This includes both a redefinition of the mechanism input for the KBE tool and the input to control the optimization including definition of design variables, the objective function and the constraint. The syntax for this input will not be discussed here.

8.5.2.2 Discussion of case results

With the selection of parameters presented in previous sections, a DoE was executed; the results of all these samples are shown in Figure 8.37 These samples were first used to create various surrogate models. The performance of the surrogate models was compared by extracting the predicted responses for the reference design. Only the nearest neighbor surrogate model type was found to give sensible results. Table 8.8 has some of the predicted values listed, together with the requirements and actual results from FEDEM.

In addition to evaluating SM types, two attempts at discovering a design that meets the requirements, with the help of optimization, were made. The first attempt used models of the responses, and the objective was calculated for each iteration, as the Cartesian distance to the requirements. The second attempt calculated the value of the objective, using the same Cartesian distance. Then a model was constructed, using the values of the objective as the samples. Both of the attempts converged, and the values for the responses are given in Table 8.8. The columns where the heading has an s subscript refer to the Reference design produced by synthesis. The $FEDEM_s$ column in the table refers to the actual values produced by FEDEM and SM_s column refers

TABLE 8.8
Stephenson 3 results [25]. (Subscripts s,r and o in headings refer to synthesis, response and objective, respectively).

Lab	SM_s	$FEDEM_s$	SM_r	$FEDEM_r$	SM_o	$FEDEM_o$	$Targ$
a_x	2.367	2.424	2.466	2.545	–	2.692	2.708
a_y	-2.260	-2.627	-2.465	-2.163	–	-2.805	-2.708
v_x	0.852	0.873	0.854	0.854	–	0.855	0.862
v_y	0.881	0.860	0.880	0.879	–	0.881	0.862
t_p	3.522	3.530	3.492	3.493	–	3.503	3.500

to the approximation. The columns in the table where the heading has an r subscript refer to Surrogate models of responses. $FEDEM_r$ refers to the actual responses at the optimum point and SM_r refers to the actual Surrogate model values at the optimum. The columns in the table where the heading has an o subscript refer to Surrogate models of the objective function. In this case the Surrogate model (SM_o) does not give individual response results and is left blank, but the $FEDEM_o$ column shows responses from FEDEM for the found optimum. The column in the table with the heading $Targ$ shows specified values from the synthesis tool. Only the velocities and acceleration at the pickup point, and the time at which the mechanism was in the pickup position, were used in the optimization.

From Table 8.8, two key observations can be made. Firstly, note that when the problem is solved by optimizing on the model of the objective, the results are closer to the target requirements, when compared with the reference design. This can be interpreted in two ways. If results from FEDEM are taken as the real behavior of the mechanism, the optimization algorithm can obtain results that are better than the analytic method used for the reference design. Since FEDEM incorporates more physics, this view has some appeal. If on the other hand, the analytical results are taken to be the true results, the optimization algorithm can get as close to the target requirements as can be expected, since the FEDEM based model is not able to give an accurate value of the true results. For both of the interpretations, the results can be said to be reasonably close, especially since the values of the responses have quite an extensive range over the design space.

9

Discussions and Conclusions

The goal of this book has been to study the available options that may be used to automate the design of mechanisms in a simulation based design process. Our approach has been to start with the positioning and orienting the joints in space, and at the same time identify the links the different joints are connecting. Based on these joint definitions the link connectivity is derived. If two joints are connected to a link it is called a binary link; if three joints are connected to the link it is call ternary; if four joints are connected to a link it is called quaternary and so on. Link geometry connections between joints in a link is constructed from a set of cross section definitions. For a binary link there is only one connection that always needs to be there. For a ternary link there are potentially three connections and all three do not necessarily need to be there; two could be enough. For a quaternary link there are 6 possible connections and you usually like to suppress some of them. We call these connections members and for higher-order links many of these connections/members need to be suppressed. More details about this can be studied in Section 4.3.

9.1 Modeling of Planar and Spatial Mechanisms

For planar mechanisms where all joint directions are the same, building link geometries is straightforward and no, or very few, special cases have to be tackled. Members are specified from a defined set of cross sections with potential varying dimensions along the members. The different members may be straight or curved specified by weighted spline points along the member. Default joint dimensions are calculated from the dimensions of the connected members with a scaling factor of 1.1. Three members in a link forming a loop will as default be connected by a surface through the governing centerlines or splines. Sharp corners where joint geometries and member geometries intersect are by default smoothed by blending in the solid geometry. Automatic FE meshing, including adjustments of element sizes, is then usually straightforward. In some cases surface thicknesses, radius for blending of sharp edges and adjustments of element sizes could be desirable. Examples of modeling of planar mechanisms in the RaMMS KBE application are shown in Section 8.1.

Modeling of 3D mechanisms in RaMMS give some additional challenges. Firstly, it is quite demanding to define spline points with weights for 3D curves; refer to the description for the double wishbone suspension example in Section 8.2.1. With the customized link designer available, this job would be much easier; see Section 7.1.2. Working with higher degree links there are challenges making the automatic generation of the link work properly. The main challenge is to pick the members that should be eliminated. In principle there is no limit for how many joints one link could have, but in practice more than five joints in a link are difficult to handle. The first link (SlewingPlate) in the offshore crane example has five joints with default 10 members; see Section 8.3.1. In mechanisms in general five joints or more for one link are quite a lot and are seldom a limitation. In connection with the simulation tool FEDEM we are also defining "free joints" and "rigid joints" for connecting structures or for defining points of interest, and in this setting being able to define many joints on a link is advantageous.

Creating a link between for instance two revolute joints close to each other, and with axis perpendicular to each other, could require a quite complex geometry being made in a CAD system, and not surprisingly is also difficult to generate automatically in RaMMS; see for instance the connecting link behind the wind turbine blades in Section 8.4.1. In the same wind turbine modeling example there was another challenge connecting the tower axially to a revolute joint at the top of the tower. In the current version of the pilot application only revolute joints and ball joints are defined with detailed joint geometry while the free joint has no local geometry defined; this is best seen in the generic example in Section 4.6.2. Other joint types like the prismatic joint", the cylindrical joint", the screw joint" and other more compound joints like the universal joint" and a gear joint" are not implemented yet, but this is only a matter of effort and time for implementation. Two revolute joints close to each other with joint axes 90 degrees (or another angle) rotated relative to each other could be regarded as a compound joint related to the universal joint.

For all joints in RaMMS, RBE2 rigid connections are generated to the FE mesh and this works very well for joints with well-defined geometries like the revolute and the ball joint. For RBE2s in free joints without a defined geometry, the connecting points to the mesh are not well defined, and the designer/analyst has little control of where these points are put. To have an optimal FE mesh, the algorithm for varying element sizes over a link needs to be revisited. An example of this challenge is the member defining the wing of the wind turbine where dimensions vary from 3.37 meters close to the hub and 0.29 meters at the tip. The smaller dimension at the tip makes the element size quite small for the whole wing structure; see Section 8.4.1. Also the thickness for shell elements should be easier to control. For the time being only 3 node shell elements and 4 node tetrahedral solid elements are used. There should also be an option to generate mesh with mid-side nodes like the 6 node shell element and the 10 node tetrahedral element.

9.2   Effective User Interaction

The user interaction to model the KBE examples in RaMMS for this book has been to edit files (in for instance Notepad++). The data defining the different examples was less than one page in total; see Tables 8.4, 8.5, 8.6 and 8.7, and therefore not requiring much typing effort. However, the work is somewhat complicated because this data is composed of several different small files that need to be coordinated to have a working model. The file containing the coordinates (coordinates.txt) is straightforward and easy to edit. The file containing constraint definitions (constraints.txt) is somewhat more challenging because what is chosen as the first and second link for each joint also effects how the geometry is generated. For instance, looking at the constraint data for the four-bar mechanism in Table 4.2, we see the crank (link 0) is the second link for both joints involved. The rocker (link 2) is also the second link in both joints involved, while the coupler link (link 1) is referred first in the corresponding joints. This will generate male joint elements on the coupler link and female joint elements on the crank and rocker links. This will put the coupler link in the back and the crank and the rocker in the front; refer to the discussion around male and female joint elements for joint generation; see Section 4.2.2. Joint direction is a vector specifying the direction of the joint axis, and is especially simple for planar mechanisms. To enter the data referred to above, and simulation input like springs, dampers, loading, etc., an interactive module is proposed; see Section 7.1.1. This module should simplify the data input through interactive graphic feedback.

Definition of link shapes could be very simple, if the same cross section could be used for all members in all links. Just one line is needed; see dataset 3 in Table 8.4. If you like to make a more advanced geometry with different cross sections, it is still quite easy with just one line for each link member specifying cross section type and cross section dimensions. However, if you are not satisfied with only straight members between joints, you need to specify points with weight factors along the member to define a spline curve that controls a geometry sweeping along the member. Specifying these spline points with weight factors is usually an iterative process. When defining spatial mechanisms this process is much more demanding; refer to the double wishbone example discussed in Section 8.2.1. To simplify this link geometry modeling a customized CAD module is proposed in Section 7.1.2. This tool is still based on the RaMMS library format and the corresponding link generation algorithm implemented in AML. A generated link geometry in RaMMS could also potentially be exported to a generic CAD system for editing before importing back into RaMMS for generating simulation input.

9.3   The Optimization Loop

In Section 8.5 some simple examples are presented in order to demonstrate how optimization can be included in mechanism design. Because we have demonstrated that KBE is used to automate all parts of mechanism design to a very large extent, the next step will be to close the design loop with automatic optimization to search for an optimal design both with respect to functionality of the design and the dimensioning with respect to stress levels and deformations. The examples referred to show that this is doable regarding the functionality of the design and this could also be extended to dimensioning with respect to design integrity. The optimization functionality implemented based on state-of-the-art optimization algorithms is generic and capable of optimizing most problems, but to make this a workable solution is a quite demanding task for most design cases that could be optimized. Selecting the design variables for the optimization is usually quite straightforward, but when it comes to the objective function and the constraints you first need to define what they should be, and next to find a mathematical representation that covers these exactly. Usually you would need some experimenting with these formulations to verify that they really give the mechanism design requested. If the search leads you into solutions that you did not expect, in many cases new constraints and even an updated objective function could be necessary to get a feasible design.

If for instance you have found a proper objective function, and the constraints necessary, and have developed mathematical formulations for these, the next challenge is to have simple user input utility that preferably does not require programming skills. It is quite a high requirement that the designer should also be a skilled programmer. One approach that has been considered is to develop templates for optimization of classes of mechanisms where some of the optimization modeling could be more or less automated. Whether this is a viable solution is not known yet, but to implement this will need much effort, and it is difficult to decide if there is enough interest to justify this effort.

To fully develop simulation optimization, ideally the complete simulation functionality modeling should be available as input to the KBE application including modeling of control system engineering. In the pilot implementation only a subset of the simulation input is made available. For automation of the manual design process excluding optimization there is not very much to gain by moving all simulation input to the KBE application. The major reduction of manual work comes from the automatic generation of the link geometry and the FE meshing; some of the simulation input may as well be given directly in the simulation tool.

As mentioned earlier, surrogate models may also be used with manual design optimization where the designer replaces the optimization algorithm.

However, also for using surrogate models manually you will have to define design parameters, objective function and constraints, so the question could then be why not also use the optimization algorithm to find an optimal solution. In the process of defining design variables, objective function and constraints, it could be advantageous to be able to manually experiment with different variations of the formulation for the objective function and the constraints in search of the best formulations.

By building surrogate models for all the responses of interest, and visualizing them in a user interface, no formulation of the objective and constraints is necessary. The engineer could investigate the behavior of the mechanism both by evaluating the SMs at specific points in the design space, and by looking at larger correlations, e.g., by plotting a design variable against a response. This would, when the SMs have been created, allow the engineer to test a new design in seconds, rather than minutes and hours. In some sense, this approach keeps the design evaluation procedure and optimization in the engineer's mind, and the optimization by the designer is connected to the model with the help of a user interface.

As discussed above, there are some challenges to utilizing optimization to a large extent in mechanism design, but having this option available for some cases will be of great advantage, and that will also influence the designer's thinking about the design iteration. The simplest approach mentioned above is to use surrogate models only for certain response variables to simplify experimentation with varying design variables and see the result without running new simulations. This requires only the definition of the design variables— usually a simple task for the designer. To utilize this approach, you typically can run a set of simulations overnight and then experiment with the design variables and see the responses without doing more simulations.

9.4   Conclusions

As discussed earlier in this chapter, we have shown in this book that much can be gained in effectiveness with mechanism design using KBE and that utilizing this potential could be a boost for using linkages in more applications opposed to using robotics, for instance for handling operations. The pilot application used for the demonstrations in this book has some limitations both regarding modeling features, for example joint types, but still covers a quite broad spectrum of mechanisms. In many cases the geometry generated by the KBE application, especially for joints, is more detailed than what is usual in simulation modeling.

The geometric variation of links possible by the modeling features of the KBE application is quite extensive, and is very much suited for parameterization. There are of course some limitations for which link geometries may be generated, but there is still potential for refinement of the modeling options both regarding link details and joint types. How much effort should be put into extending more detailed simulation input into the KBE application depends on to what extent automatic optimization will be utilized.

As discussed above, one of the most straightforward features of the optimization tools made available could be to use Surrogate Models for responses in the simulation, stepping the design variables systematically over the feasible ranges, generating mechanism designs and simulation input for the different combinations using the KBE application and then running the simulations in batch. Following this, the designer may in a very effective way interactively experiment with changing design variables and see the simulation responses without running new simulations.

The pilot implementation is using simple files as input that are edited manually. Chapter 7 proposes tools for modeling this input interactively. Especially the link designer described in Section 7.1.2 will simplify the modeling of three-dimensional mechanisms very much.

The KBE pilot is already a very affective tool for modeling a wide range of mechanisms for simulation with a tremendous reduction in manual modeling time used. Automatic optimization is not in a form recommended for use by anyone outside the development team, but there are efforts to make these options available also for a wider community.

References

[1] R.L. Ackoff. From data to wisdom. *Journal of Applied Systems Analysis*, Vol. 16:3–9, 1989.

[2] Siemens AG. *NX Nastran User's Guide*. Siemens Product Lifecycle Management Software Inc, 2014.

[3] FEDEM Technology AS. *The FEDEM Simulation System*. The FEDEM Technology Company, Trondheim.

[4] M.C.C. Bampton and R.R. Craig. Coupling of substructures for dynamic analyses. *AIAA Journal*, 6(7):1313–1319, 1968.

[5] P.G. Bergan and M.K. Nygard. Nonlinear shell analysis using free formulation finite elements. *Finite Element Methods for Nonlinear Problems*, pages 317–338, 1986.

[6] Bertold Bongardt. Shethuicker convention revisited. *Mechanism and Machine Theory*, 69(0):200–229, 2013.

[7] C.B. Chapman and M. Pinfold. Design engineering: A need to rethink the solution using knowledge based engineering. *Knowledge-Based Systems*, 12(56):257–267, 1999.

[8] Craig B. Chapman and Martyn Pinfold. The application of a knowledge based engineering approach to the rapid design and analysis of an automotive structure. *Advances in Engineering Software*, 32(12):903–912, 2001.

[9] J.J. Chung and G.M. Hulbert. A time integration algorithm for structural dynamics with improved numerical dissipation: The generalized-α method. *J. Appl. Mech.*, 60:371–375, 1993.

[10] Thor Christian Coward. *User Interface Concepts for Mechanism Modelling in the RaMMS KBE System*. Master's Thesis, Norwegian University of Science and Technology, NTNU, 2017.

[11] DaimlerChrysler. MOKA, Task 4.3, Final Synthesis, EU ESPRIT MOKA project, EP 25418, 2000.

[12] J. Denavit and R.S. Hartenberg. *Kinematic Synthesis of Linkages*. McGraw-Hill, 1964.

[13] Dov Dori. *Model-Based Systems Engineering with OPM and SysML.* Springer, New York, 2015.

[14] Guus Schreiber et al. *Knowledge Engineering and Management—The CommonKADS Methodology.* The MIT Press, 2000. 0-262-19300-0, 2000.

[15] Clark C. Evans. The Official YAML Web Site. URL: http://yaml.org/.

[16] C.A. Felippa and B. Haugen. A unified formulation of small-strain corotational finite elements: I. theory. *Computer Methods in Applied Mechanics and Engineering*, 194(2124):2285–2335, 2005. Computational Methods for Shells.

[17] L. Frøyd and O.G. Dahlhaug. A conceptual design method for parametric study of offshore wind turbines. *Proc. of the ASME 2011 30th Int. Conf. on Offshore and Arctic Engineering*, OMAE 2011. Rotterdam.

[18] Philip E. Gill, Walter Murray, and Michael A. Saunders. *SNOPT: An SQP Algorithm for Large-Scale Constrained Optimization.* Society for Industrial and Applied Mathematics, 2005.

[19] Friedel Hartmann. *The Mathematical Foundation of Structural Mechanics.* Springer-Verlag, 1985.

[20] W. C. Hurty. Vibrations of structural systems by component mode synthesis. *Journal of the Engineering Mechanics Division*, 86(4), 1960.

[21] John T Hwang. *A Modular Approach to Large-Scale Design Optimization of Aerospace Systems.* PhD dissertation, University of Michigan, 2015.

[22] TechnoSoft Inc. *AML Basic Training Manual.* TechnoSoft Inc., v.3.06 edition, TechnoSoft Inc., Ohio, 2007.

[23] P. Klein, J. Lutzenberger, K. Kristensen, and G. Iversen. K-brief and context driven access: Providing context related information to product developers in high quality. *Engineering, Technology and Innovation (ICE), 2014 International ICE Conference on 23-25 June 2014*, 2014.

[24] Anders K. Kristiansen and Eivind Kristoffersen. *Automating Tasks in the Design Loop for Mechanism Design.* Master's Thesis, Norwegian University of Science and Technology, NTNU, 2016.

[25] Arnt Underhaug Lima. *Optimizing KBE Generated Mechanisms.* Master's Thesis, Norwegian University of Science and Technology, NTNU, 2017.

[26] LinkedDesign. The LinkedDesign EU project.

[27] P.J. Lovett, A. Ingram, and C.N. Bancroft. Knowledge-based engineering for {SMEs} a methodology. *Journal of Materials Processing Technology*, 107(13):384–389, 2000.

[28] David G. Luenberger. *Linear and Nonlinear Programming, Second Edition.* Kluwer, Boston, 2003.

[29] Ivar Marthinusen. *The Acquisition and Codification of Knowledge for Knowledge-Based Engineering.* PhD dissertation, Norwegian University of Science and Technology (NTNU), 2016.

[30] Nick R. Milton. *Knowledge Acquisition in Practice: A Step-by-Step Guide.* Springer, London, 2007.

[31] A. Myklebust and D. Tesar. The analytical synthesis of complex mechanisms for combinations of specified geometric or time derivatives up to the fourth order. *Journal of Engineering for Industry, Trans. ASME,* 1975.

[32] A. Myklebust and D. Tesar. Application and design equations for synthesis of complex mechanisms for combinations of higher order time and geometric derivative specifications. *ASME Paper No. 74-DET-77, Design Engineering Technical Conference,* New York, 1974.

[33] Arvid Myklebust. *Synthesis of Multi-Link Mechanisms for Dynamic Specifications.* PhD dissertation, University of Florida, 1974.

[34] Jorge Nocedal and Stephen J. Wright. *Numerical Optimization.* Springer, New York, 2006.

[35] R. Norton. *Design of Machinery (McGraw-Hill Series in Mechanical Engineering).* McGraw-Hill Education, 2011.

[36] B. Nour-Omid and C.C. Rankin. Finite rotation analysis and consistent linearization using projectors. *Computer Methods in Applied Mechanics and Engineering,* 93(3):353–384, 1991.

[37] OMG. *Systems Modeling Language (SysML).* The Object Management Group, Needham, MA 02494 USA.

[38] OpenMDAO. *Multidisciplinary Design, Analysis, and Optimization (MDAO).* NASA Glenn Research Center.

[39] J.A. Penoyer, G. Burnett, D.J. Fawcett, and S.-Y. Liou. Knowledge based product life cycle systems: principles of integration of kbe and c3p. *Computer-Aided Design,* Vol. 32(Issues 56, May 2000):311–320, 1999.

[40] Martyn Pinfold and Craig Chapman. The application of KBE techniques to the FE model creation of an automotive body structure. *Computers in Industry,* 44(1):1–10, 2001.

[41] Martyn Pinfols, Craig Chapman, and Steve Preston. Knowledge acquisition and documentation for the development of a KBE system for automated FE analysis. *International Journal of Knowledge Management Studies,* Vol. 2(No. 2), 2008.

[42] Steve Preston and Craig Chapman. Knowledge acquisition for knowledge-based engineering systems. *International Journal of Information Technology and Management*, Vol. 4(No. 1), 2005.

[43] Dante Pugliese and Giorgio Colombo. D6.3 Rule Interchange Format Standardization Document.

[44] C.C. Rankin and B. Nour-Omid. The use of projectors to improve finite element performance. *Computers & Structures*, 30(1):257–267, 1988.

[45] F. Reuleaux. *The Kinematics of Machinery: Outlines of a Theory of Machines*. Macmillan, London, 1876.

[46] Gianfranco La Rocca. Knowledge based engineering: Between {AI} and {CAD}. review of a language based technology to support engineering design. *Advanced Engineering Informatics*, 26(2):159–179, 2012.

[47] Susana Rojas-Labanda and Mathias Stolpe. *Benchmarking optimization solvers for structural topology optimization*. Struct Multidisc Optim 52 (2015).

[48] Phillip Sainter, Keith Oldham, and Andrew Larkin. *Achieving benefits from knowledge-based engineering systems in the longer term as well as in the short term*. Knowledge Engineering and Management Centre, Coventry University, Priory Street, Coventry, 2000.

[49] Nigel Shadbolt and Mike Burton. The empirical study of knowledge elicitation techniques. *SIGART Newsletter. April 1989, 108, pp. 15 - 18*, Vol. April 1989(No. 108):15–18, 1989.

[50] P. Sheth and J.J. Uicker. IMP (Integrated Mechanisms Program), A computer-aided design analysis system for mechanisms and linkage. *J. Manuf. Sci. Eng.*, 94(2):454–464, 1972.

[51] Ole Ivar Sivertsen. *Virtual Testing of Mechanical Systems, Theories and Techniques*. Swets & Zeitlinger, Trondheim, Norway, 2001.

[52] Rasmus Korvald Skare. *Mechanism Parametrization, Modeling and FE-Meshing*. Master's Thesis, Norwegian University of Science and Technology, NTNU, 2015.

[53] J. Sobieski and O. Storaasli. Computing at the speed of thought. Feature article in *Aerospace America*. *The Monthly Journal of the American Institute of Aeronautics and Astronautics*, 2004.

[54] Jaroslaw Sobieszczanski-Sobieski, Alan Morris, Michel J.L. van Tooren, Gianfranco La Rocca, and Wen Yao. *Multidisciplinary Design Optimization Supported by Knowledge Based Engineering*. Ringgold Inc., John Wiley & Sons, Ltd., UK, 2015.

[55] Siemens PLM Software. *Parasolid, 3D Geometric Modeling Engine.* Plano, TX 75024, USA.

[56] Bernhard Specht. *Optimization Theory and Software Requirement Document.* ESPRIT 5524 Project,D2203, Dornier, Germany, 1992.

[57] Bernhard Specht. *Sensitivity Theory and Software Requirement Document.* ESPRIT 5524 Project,D2201, Dornier, Germany, 1992.

[58] Melody Stokes. Managing engineering knowledge—MOKA: Methodology for knowledge based engineering applications. *Professional Engineering Publishing Limited, ISBN: 1860582958*, 2001.

[59] Krister Svanberg. The method of moving asymptotes—a new method for structural optimization. *Structural and Multidisciplinary Optimization*, Vol. 48(No. 5):859–875, 1987.

[60] TechnoSoft Inc. *Adaptive Modeling Language (AML).* The TechnoSoft Inc. Company, Cincinnati.

[61] Sigurd D. Trier. *Design Optimization of Flexible Multibody Systems.* PhD dissertation, Norwegian University of Science and Technology (NTNU), 2001.

[62] J. J. Uicker, G. R. Pennock, J. E. Shigley, and J. M. McCarthy. *Theory of Machines and Mechanisms.* Oxford University Press, 2003.

[63] John J. Uicker, Bahram Ravani, and Pradip N. Sheth. *Matrix Methods in the Design Analysis of Mechanisms and Multibody Systems.* Cambridge University Press, 2013.

[64] Wim J.C. Verhagen, Pablo Bermell-Garcia, Reinier E.C. van Dijk, and Richard Curran. A critical review of knowledge-based engineering: An identification of research challenges. *Advanced Engineering Informatics*, 26(1):5–15, 2012. Network and Supply Chain System Integration for Mass Customization and Sustainable Behavior.

[65] Andreas Wächter and Lorenz T. Biegler. On the implementation of an interiorpoint filter line-search algorithm for large-scale nonlinear programming. *Mathematical Programming*, Vol. 106(No. 1):25–57, 2006.

[66] Z. Xu Z. Lyu and J.R.R.A. Martins. *Benchmarking Optimization Algorithms for Wing Aerodynamic Design Optimization.* The Eighth International Conference on Computational Fluid Dynamics (2014).

Index

A

A3 knowledge briefs, 20, 22

Actuators, 3, 109, 133

Adaptive Modeling Language (AML), 93

    Boolean geometry, 95–96, 97*t*, 98*t*

    crane model case, 133, 134*f*

    extended KBE programming, 112

    finite element data for mechanism links, 64–65

    graphic environment, 66

    KBE development framework, 93–94

    pilot application, *See* Rapid Mechanism Modeling System

    rule interchange format, 20–21

    wind turbine case, 141*f*, 142*f*

Advisory KBE systems, 9

Algorithms and design optimization problems, 80

Alpha (α) method, generalized, 49–51

AMETank, 25

AML, *See* Adaptive Modeling Language

AMOpt module, 94

Artificial intelligence (AI), 7, 13

Automated design, 1, 161, 164

    closing the design loop, 164

    example mechanism demonstration, 66–74, *See also* Automated design cases; Dynamic simulation,

    example mechanism demonstration

    finite element data for mechanism links, 64–65

    knowledge acquisition for KBE application development, 24–27, *See also* Knowledge acquisition

    knowledge-based engineering, 2–3, *See also* KBE applications; Knowledge-based engineering

    mechanism system data, 65

    optimization, 2, *See also* Design optimization

    pilot application, *See* Rapid Mechanism Modeling System

    simulation tool, *See* FEDEM

Automated design cases, 113

    cranes and robots, 132–138

    four-bar mechanisms, 113–120

    linkages, 113–124

    six-link mechanisms, 116, 119*f*, 120, 122*f*, 123*f*, 124*f*

    suspension systems, 120, 125–132

    wind turbines, 138–146, 162

Automated design environment, 93

    Boolean geometry, 95–96, 97*t*, 98*t*

    design parameters and optimization, 110–112

    generative design, 95

    KBE development framework, 93–94

user interface, *See* User
    interaction
See also KBE applications;
    Rapid Mechanism Modeling
    System
Automatic FE meshing, 96, 99
"Automation in design," 2, *See also*
    Automated design

B

Ball joints, 66, 70, 125
Barrier function methods, 82
Bzier curves, 61
Binary link geometry, 95–96, 161
    difference class in AML, 98*t*
    union class in AML, 97*t*
Blending of sharp edges, 71, 72, 106,
    129, 161
Boolean geometry, 95–96
B-splines, *See* NURBS
Bulk Data Format (BDF), 64, 92, 94,
    103

C

Caching, 10–11
Camber angle, 125
Case-based reasoning (CBR), 14
Charts, 17
CommonKADS, 16
Communication rules, 12
Complete graph problem, 60
Component mode synthesis (CMS)
    transformation, 35, 65
Computer aided design (CAD), 2–3,
    7
    extended KBE programming,
        112, 163
    modeling link shapes, 64, 163
    PARASOLID kernel, 95, 112
Concept mapping, 19
Conditional expressions, 11
Configuration selection rules, 11–12
Conjugate gradient methods, 81
Constant velocity joints, 41–42

Constrained optimization methods,
    82, 148, 151
Constraints, design optimization
    problem formulation, 78
Constraints, joints as, *See* Joints
Constraints, multipoint (MPC), 44,
    71, *See also* RBE2
    connections
Continuous design variables, 77
Control engineering modeling,
    109–110
Control systems
    dynamic simulation, 3
    mechanism model system data,
        65
Convergence tolerance simulation
    parameters, 108
Convex approximation methods,
    82–83
Convex functions and problems, 80
Corotational finite element
    formulation, 33
Coupler link, 66, 70*f*
Craig-Bampton modes, 35
Cranes, 132–138
Critical decision making (CDM), 19
Cylindrical joints, 41, 43, 162

D

Dampers
    automated design cases, 114
    dynamic simulation, 3
    entering/editing mechanism
        library format, 104
    example mechanism
        demonstration, 66
    generic library format for
        mechanisms, 59–60
    mechanism model system data,
        65
Damping forces, 37–38, 47
Deformational forces, 47
Degrees of freedom (DOFs)
    joint generic library format, 58
    mechanism static modes, 34

Dependency tracking, 10–11
Design automation, *See* Automated
design
Design automation cases, *See*
Automated design cases
Design automation environment, *See*
Automated design
environment; FEDEM
Design optimization
about optimization techniques,
4–5
automated optimization and
design parameters, 110–112
direct optimization, 111
dynamic performance, 86–87
finite difference method, 88
flexible multibody systems,
87–92
four-bar mechanism case,
146–153
fundamental terms and
concepts, 79–81
global optimization methods, 83
KBE development framework,
93–94
manual approaches, 111–112
methods for constrained
optimization, 82–83, 148,
151
methods for unconstrained
optimization, 81, 148
multicriteria or multiobjective
optimization, 78, 84–85
multidisciplinary optimization,
83–84
Pareto optimization, 84–85
problem formulation, 77–79
sensitivities,objectives, and
constraints, 87–89
six-link mechanism case,
153–159
standard problem, 79
surrogate models, 89–90, 92,
111, 151–153, 164–165, 166
work-flow improvement, 15, 153

See also Automated design
environment
Design sensitivities, 87–89
Design variable, optimization
problem formulation, 77
Design work-flow improvement, 15
Difference operation, 95
Direct differentiation method, 88
Direct optimization, 111
Discrete design variables, 77
Door model, 132–133
Double wishbone suspension, 108,
125–132
Dynamic performance optimization,
86–87
Dynamic simulation, 1, 3–4
corotational formulation, 33
description of motion, 29–32
design automation environment,
See Automated design
environment
example mechanism
demonstration, 66–76
FE meshing, 71–74
importing into FEDEM,
74–76
joint definition input, 66–67
link shape definition input,
67–71
FE modeling of joints, 39–40
FE modeling tool, *See* FEDEM
finite element (FE) theory, 32
fixed interface modes, 35
flexible multibody systems,
87–92
generic mechanism modeling
framework, *See* Generic
mechanism modeling
framework
joint specializations, 40–44, *See
also* Joints; *specific types*
links as substructures and super
elements, 34–38
mechanism analysis, 34

mechanism library format, 103,
    See also User interaction
model reduction, 34
multipoint constraints, 44–47
reduced system, 35–37
static mechanism modes, 34
structural damping, 37–38
substructure dynamic equation
    of motion, 36
time integration for nonlinear
    dynamics, 47–53

E
Euler's rotation theorem, 31

F
Feasible set, 79
FE-based dynamic simulation, *See*
    Dynamic simulation;
    FEDEM
FEDEM, 64, 92, 93
    crane model case, 133, 136*f*,
        137*f*, 138, 139–140*f*
    double wishbone suspension
        model, 125, 128–132*f*
    example mechanism
        demonstration, 66, 74–76,
        See also Dynamic
        simulation, example
        mechanism demonstration
    four-bar mechanism case, 114,
        117*f*, 148
    six-link mechanism case, 120,
        154, 158–159
    spatial mechanism modeling,
        162
    system input file, 64–65
    user interface, 103, 110, *See also*
        User interaction
    wind turbine case, 144*f*, 145*f*,
        146
Finite difference method, 88, 153
Finite element (FE) analysis, KBE
    application development
    framework, 93–94

Finite element (FE) mesh
        generation, 3
    blending of sharp edges, 71, 72,
        106, 129, 161
    design automation environment,
        96, 99
    KBE applications, 4
    See also Finite element (FE)
        modeling of joints
Finite element (FE) modeling of
        joints, 39–40, 71–74
Finite element (FE) simulation tool,
        See FEDEM
Finite element (FE) theory, 32
    corotational formulation, 33
    virtual work equation, 32
Finite element-based dynamic
        simulation, *See* Dynamic
        simulation
Finite element data for mechanism
        links, 64–65
Fixed interface dynamic modes, 35
Fixed joints, 27, 40
    crane and robot automated
        design case, 133
    example mechanism
        demonstration, 66, 67
    modeling in FEDEM, 74
Flexible multibody systems (FMBS)
    multidisciplinary design
        optimization, 83–84
    optimization tool
        (OpenMDAO), 90–92
    optimizing, 87–92
Flexible multipoint constraints,
        45–47
Focused discussion, 18
Four-bar mechanism, 57*f*
    automated design cases,
        114–116
    example mechanism
        demonstration, 66, 68*f*, 69*t*,
        75*f*
    optimization demonstration,
        146–153

Frame notation, 55
Free joints, 27
  example mechanism
    demonstration, 66, 67, 70,
    71
  modeling in FEDEM, 74
  spatial mechanism modeling,
    162
Function modeling simulation input,
  109

G
GDL (General-purpose Declarative
    Language), 7
Gear joint, 162
Gear transmission, 44
Generalized α method, 49–51
Generative KBE systems, 9
Generic mechanism modeling
    framework, 55
  automated generation of
    simulation input, 64–65
  default link and joint shapes,
    60–63
  frame and pose notation, 55
  library format, 56–60
    constraints, 56–58
    link shapes, 58–59
    node positions, 56
    springs and dampers, 59–60
  Sheth-Uicker formulation, 55–56
  See also Dynamic simulation
Genetic algorithms (GA), 83, 90
Geometric manipulation rules, 11
Ghost reference, 33
Global convergence of algorithms, 80
Global minimum points, 79
Global optimization methods, 83
Global solutions, 79
Graphical user interface, *See* User
    interaction
Grid techniques, 19
Guided random search techniques
    (GRST), 83

H
Hessian matrix, 80
Higher pairs, 25, 26
Hocken's linkage, 57*f*

I
ICARE forms, 17
IDL (ICAD Design Language), 7
Inertia forces, 47–48
Innovative KBE systems, 9–10
Integration methods, 47–53
Intent! 7
Interior point methods, 82, 90
Intersection operation, 95
IPOPT, 90
Iterative algorithms, 80

J
Joint list, 56–58
Joint modeling window, 104, 105*f*
Joints, 1, 161
  assembly, 63
  automated design approach, 161
  complete graph problem, 60
  crane and robot automated
    design case, 132–133
  definition, 25
  degrees of freedom variation, 40
  door model, 132–133
  dynamic simulation, 3
  entering/editing mechanism
    library format, 104, 163
  example mechanism
    demonstration
    definition input, 66–67
    link shape definition input, 70
  FE modeling, 39–40
  finding dimensions, 63
  generic library format for
    mechanisms, 56–58
  generic mechanism modeling
    default shapes, 60–63
  geometry representation, 63
  kinematic pairs, 25

main frames and sub-frames, 63,
    70
mechanism model system data,
    65
mechanism theory for KBE
    development, 25–27
mechanism topology, 25
modeling planar mechanisms,
    161
multipoint constraints, 44, 71,
    See also RBE2 connections
positioning in automated design
    cases, 116
rigid body motion description,
    30
spatial mechanism modeling,
    162
specializations, 40–44, *See also*
    specific types
transmissions between variables,
    43–44
See also specific types

K
KBE applications, 3
    automated design, *See*
        Automated design;
        Automated design cases
    classical mechanism theory for
        developing, 25–27
    developing, 23–24
    development framework, 93–94,
        164
    dynamic simulation, 4
    example mechanism
        demonstration, 66–76, *See*
        also Dynamic simulation,
        example mechanism
        demonstration
    extended KBE programming,
        112, 163, *See also*
        Computer aided design
    implementation issues, 23–24
    knowledge acquisition for
        developing, 21–22, 24–27,

    See also Knowledge
        acquisition
    main rules, 11–12
    optimizing flexible multibody
        systems, 87–92
    pilot application, *See* Rapid
        Mechanism Modeling
        System
    pilot application limitations,
        164–165
    programming languages, *See*
        KBE programming
        languages
    user interface, *See* User
        interaction
    See also Adaptive Modeling
        Language; Automated
        design environment; Rapid
        Mechanism Modeling
        System
KBE frameworks, 9
KBE life-cycle, 16–17
KBE programming languages, 7–8,
    10–12, 93
    KBE development framework,
        93–94
    rule interchange format, 20–21
    See also Adaptive Modeling
        Language
KBE selection systems, 10
Keypoint modeling menus,
    103–104
Kinematic pairs, 25–26
Knowledge, 8–9, 15
Knowledge, tacit, 13, 23
Knowledge acquisition, 12–22
    automation (KBE) into
        mechanism design, 24–27
    components, 14
    desirable expert characteristics,
        23–24
    elicitation from experts, 14,
        18–19
    KBE application development,
        21–22

potential difficulties, 13
tools and methodologies, 16–22
Knowledge analysis, 14
Knowledge base, 7, 12, 15
Knowledge-based engineering
    (KBE), 2–3, 13–14
  AI and, 7, 13
  CAD systems and, 7
  concepts and classification, 8–10
  definition, 9
  knowledge acquisition for,
    12–22, *See also* Knowledge
    acquisition
  life-cycle, 16–17
  programming languages, *See*
    KBE programming
    languages
  See also KBE applications
Knowledge-based reasoning, 14
Knowledge briefs (K-Briefs), 20, 22
Knowledge elicitation, 14, 18–19, *See*
    also Knowledge acquisition
Knowledge modeling, 14
Knowledge objects, 14
Knowledge pyramid, 8*f*
Knowledge representation, 14

L
Laddering techniques, 19
Lagrangian methods, 82
Life-cycle, KBE, 16–17
Limited information task, 19
Linear programming methods, 82
Link editing module, 106–108
Links, 1, 25, 113, 161
  assembly, 63
  automated design approach, 161
  automated design cases, 113–124
  automatic FE meshing, 96
  complete graph problem, 60
  definition, 25
  entering/editing mechanism
    library format, 104,
    106–108, 163

example mechanism
    demonstration, 67–71
extended modeling of shapes,
    64, 163
finite element data for, 64–65
generic library format for
    mechanisms, 58–59
generic mechanism modeling
    default shapes, 60–63
mechanism model system data,
    65
mechanism topology, 25
modeling planar mechanisms,
    161
optimization demonstration,
    146–159
relative motion description,
    39–40
substructures and super
    elements, 34–38
surfaces, 62
sweeps, 61–62
theory for KBE application
    development, 27
Links degree, 25
Link Shape list, 58–59
Lisp, 7, 93
Load modeling menu, 104, 107
Loads, generic library format for
    mechanisms, 60
Local minimum points, 79
Logic rules, 11
Lower pairs, 25, 26, 27

M
Main-frame for a joint, 63, 70
Math rules, 11
Matrix and matrix vector notation,
    29
McPherson strut, 120
Mechanism design automation, *See*
    Automated design
Mechanism design optimization, *See*
    Design optimization
Mechanism library format, entering

and editing, 101–108, *See also* User interaction

Mechanism modeling, *See* Dynamic simulation; Generic mechanism modeling framework

Mechanisms, 1, 113
definition, 25
links as substructures and super elements, 34–38
planar mechanism modeling, 161
spatial mechanism modeling, 162
theory for KBE application development, 25–27
topology, 25
See also specific components

MECSYN synthesis program, 116, 121*f*

Menus, 103–106

Method of steepest descent, 81

Minimum points, 79

Minimum potential energy, 32

MML (MOKA Modeling Language), 18

Model reduction, 34

MOKA, 16–18

MOKA Modeling Language (MML), 18

Motion description for dynamic simulation, 29–32, 36

Multicriteria or multiobjective optimization, 78, 84–85

Multidisciplinary design optimization (MDO), 83–84

Multimaster joints, 42–43

Multiplier methods for constrained optimization, 82

Multipoint constraints (MPCs), 71, 96, 99, *See also* RBE2 connections

N

NASTRAN Bulk Data Format (BDF), 64, 92, 94, 103

Newmark integration, 48–51

Newton's method, 81

Newton's second law, 47

Node positions, generic library format for mechanisms, 56

Nonlinear dynamics, time integration methods for, 47–53

Nonlinear problem optimization algorithms, 90

Non-uniform rational B-spline (NURBS), 59, 61

Notepad++, 101, 163

NURBS (non-uniform rational B-spline), 59, 61

O

Objective function, 4
automated optimization and FE analysis, 94
constrained optimization methods, 82
dynamic performance optimization, 86–87
multiobjective and Pareto optimization, 84–85
optimization problem formulation, 78
surrogate models, 89
unconstrained optimization methods, 81
See also Design optimization

Object oriented (OO) programming languages, 7–8, 10, 93–94

OpenMDAO, 90–92, 152

OPM (Object Process Methodology) language, 21

Optimality conditions, 79–80

Optimization problem formulation, 77–79

Optimization techniques, 2, 4–5, *See also* Design optimization

P

Parallel processing, 89–90, 111

PARASOLID, 95, 112

Pareto optimization, 84–85
Penalty function methods, 82
Performance optimization, 86–87
Planar mechanism modeling, 161
Pose notation, 55
Positive definite and semi-definite
    Hessian, 80
Potential energy, minimum, 32
Prismatic joints, 41, 42, 43, 162
Problem formulation for design
    optimization, 77–79
Proportional damping, 37
Protocol analysis, 18

Q
Quadratic penalty functions, 82
Quasi-Newton method, 81, 82
Quaternary link, 161

R
Rack and pinion, 44
RaMMS, *See* Rapid Mechanism
    Modeling System
Rapid Mechanism Modeling System
    (RaMMS), 101
  cranes and robots case, 133
  design parameters and
      automated optimization,
      110–112
  double wishbone suspension
      model, 125, 129*f*
  entering/editing mechanism
      library format, 101–108, *See
      also* User interaction
  extended KBE programming,
      112, 163
  four-bar mechanism case, 120
  inputs not modeled for RaMMS
      KBE pilot, 108–110
  pilot application limitations,
      164–165
  planar mechanism modeling, 161
  six-link mechanism case, 120
  spatial mechanism modeling,
      162

  user interface, 163, *See also* User
      interaction
  wind turbine case, 146
  See also Automated design cases
Raster angle, 125
Rayleigh-damping, 37–38
RBE2 connections, 71, 74, 96, 99,
    103, 133, 162
Response surfaces (RS), 89
Reuse and KBE applications, 23
Revolute joints, 27, 40–41
  cranes and robots automated
      design case, 133
  double wishbone suspension, 125
  example mechanism
      demonstration, 66, 67, 70,
      72, 73*f*
  spatial mechanism modeling,
      162
Rigid body motion description,
    29–31
Rigid joints, *See* Fixed joints
Rigid multipoint constraints, 44, 71,
    See also RBE2 connections
Robots, 113–114, 132
Rocker link, 67, 71
Rodrigues parametrization of
    rotations, 31–32
Rotation matrix or tensor, 31
Rotation motion description, 31–32
Rule-based reasoning (RBR), 14
Rule interchange format (RIF),
    19–21
Rules in KBE applications, 11–12

S
Screw joints, 44, 162
Selection KBE systems, 10
Sensitivity equations, 88
Sequential linear programming
    methods, 82
Sequential quadratic programming
    (SQP) methods, 82, 90, 153
Shadow element, 33

Sharp edges, blending of, 71, 72, 106,
        129, 161
Sheth-Uicker (SU) formulation,
        55–56
Ship design process, 15–16
Simulation, 1, *See also* Dynamic
        simulation
Single-level design optimization
        (S-LDO), 84
Six-link mechanisms, 116, 119*f*, 120
    input data for, 122*f*
    mesh, 123*f*
    optimization demonstration,
        153–159
    See also Stephenson six-link
        mechanisms
SLSQP, 90
SNOPT, 90
Spatial mechanism modeling, 162
Spline curve representation, 61, *See
        also* NURBS
Springs
    automated design cases, 114
    dynamic simulation, 3
    entering/editing mechanism
        library format, 104
    example mechanism
        demonstration, 66
    gencric library format for
        mechanisms, 59–60
    mechanism model system data,
        65
SQ, 15
Static mode mechanism simulation,
        34
Steepest descent, method of, 81
STEP files, 64, 93, 112
Stephenson six-link mechanisms,
        116, 119*f*
    coupler position and velocity
        curves, 124*f*
    input data for, 122*f*
    meshed, 123*f*
    optimization demonstration,
        153–159

Structural damping, 37–38
Sub-frames, 63, 70
Substructure dynamic equation of
        motion, 36
Substructure matrices, 65
Superelement damping matrix, 37
Superelement displacements, 35–37
Super element matrices, 65
Super nodes, 96
Surface generation, 108
Surface representations, 62
Surrogate models (SMs), 89–90, 92,
        111, 151–153, 159, 164–165,
        166
Suspension systems, 108, 120,
        125–132
Sweeps, 61–62
SysML, 20
System input file, 64–65

T
Tacit knowledge, 13, 23
Teach back, 18
Tensor notation, 29
Ternary link, 161
Text editor, 101, 163
Three-dimensional (3D) mechanism
        modeling, 162
Time dependent response vector, 86
Time integration methods, 47–53
Time integration simulation
        parameters, 108
Time step simulation parameters,
        108
Toe-out or toe-in, 125
Topology of mechanisms, 25
Topology rules, 11–12
Transformation methods for
        constrained optimization,
        82
Transmissions between joint
        variables, 43–44
Triadic elicitation, 19
Twenty questions, 19

U
UML (Unified Modeling Language), 18
Unconstrained optimization methods, 82–83, 148
Unified Modeling Language (UML), 18
Union operation, 95
Universal joint, 42*f*, 162
User interaction, 101, 163
    design parameters and automated optimization, 110–112
    inputs not modeled for RaMMS KBE pilot, 108–110
    KBE application development cycle and, 24
    mechanism library format, 101–103
    link editing module, 106–108
    menus, 103–106
    using text editor, 101, 163

V
Vector notation, 29
Vibration mode simulation parameters, 108–109
Virtual work equation, 32

W
Watt six-link mechanisms, 116
Weighted average multipoint constraints, 45–47
Wind turbines, 138–146, 162
Wishbone suspension, 108, 125–132
Work-flow improvement, 15, 153

Y
YAML, 159